普通高等教育"十一五"国家级规划教材
高等职业院校精品教材系列

电子技术专业英语
（第4版）

朱一纶　编著

电子工业出版社
Publishing House of Electronics Industry
北京·BEIJING

内 容 简 介

本书涉及电子元器件、集成电路、基本放大电路、数字电路等课程的基本知识，旨在逐步提高学生的阅读、理解和翻译电子技术专业书刊资料的能力，为学生今后能够以英语为工具，获取和交流专业技术信息打下良好的基础。

本书注重选用各种不同类型的资料，有教材、说明书、广告、科普资料，并附有较多的插图，以达到比较好的教学效果，同时可以拓宽学生的知识面。本书还简单介绍了翻译知识，并配有适量的练习与扩展阅读材料，可供教师在教学中选用和学生自学。

本书适合作为高等工业专科学校教育和高等职业教育的应用电子技术专业的教材，也可作为其他电类专业学生的参考书。

未经许可，不得以任何方式复制或抄袭本书之部分或全部内容。
版权所有，侵权必究。

图书在版编目(CIP)数据

电子技术专业英语：中英对照/朱一纶编著．—4 版．—北京：电子工业出版社，2015.5（2022.6 重印）
高等职业院校精品教材系列

ISBN 978-7-121-25995-1

Ⅰ．①电… Ⅱ．①朱… Ⅲ．①电子技术—英语—高等职业教育—教材 Ⅳ．①H31

中国版本图书馆 CIP 数据核字（2015）第 094001 号

策　　划：	陈晓明
责任编辑：	郭乃明　　特约编辑：范　丽
印　　刷：	北京盛通商印快线网络科技有限公司
装　　订：	北京盛通商印快线网络科技有限公司
出版发行：	电子工业出版社
	北京市海淀区万寿路 173 信箱　邮编　100036
开　　本：	787×1 092　1/16　印张：17.75　字数：454 千字
版　　次：	2003 年 7 月第 1 版
	2015 年 5 月第 4 版
印　　次：	2022 年 6 月第 13 次印刷
定　　价：	46.00 元

凡所购买电子工业出版社图书有缺损问题，请向购买书店调换。若书店售缺，请与本社发行部联系，联系及邮购电话：(010) 88254888。

质量投诉请发邮件至 zlts@phei.com.cn，盗版侵权举报请发邮件至 dbqq@phei.com.cn。
服务热线：(010) 88258888。

前　言

　　专业英语的教学目的是指导学生阅读与自己专业相关的英语书刊和文选，使学生能以英语为工具，获取与专业相关的信息，本书的教学对象是高等工业专科学校和高等职业院校应用电子技术专业的学生，也可以作为其他电类专业学生的教学参考书。

　　本书自第 1 版（2003 年）、第 2 版（2006 年）、第 3 版（2009 年）出版以来，承蒙广大教师、学生及读者的厚爱，被许多学校选为电类专业英语的教材，共计印刷了 30 余次之多。

　　在此期间，作者不断得到读者从不同角度提出的鼓励、希望和建议，为此我们在第 3 版的基础上进行了修订。考虑到教学的要求和连贯性，修订后的本书仍然保留了原第 3 版的优点，即：

　　1．涉及的知识面广，选用了很多介绍实际的电子技术应用的内容，使教材具有较好的可读性，不仅作为专业英语的学习，还可以了解到一些常用电子电气设备的工作原理。

　　2．本书是以学习英语的科技译文、翻译知识、英语科技论文的写作、英语说明书的写作等基础知识的讲授为主线，而不是简单的学习英语单词；鉴于第 3 版的主要内容涉及到电子技术课程的许多基础知识，且反响很好，考虑到老师们教学的连贯性，故本书仍然沿用了第 3 版的内容。

　　3．此次修订中，改正了第 3 版中的一些错误之处。

　　4．本书仍然使用第 3 版原有的有声读物，即由发音纯正的英语系国家人士朗读教材中单元课文，以利于教师的教学和学生的自学，也可使学生锻炼口语，增加学习的趣味。

　　5．本书配有有声读物，可在电子工业出版社的华信教育资源网上免费下载。

　　本书由南京金陵科技学院的朱一纶编著，吴岱曦参加了资料整理、文字录入及电子教案的制作，南京金陵科技学院的吴彪等也帮助进行了校对等工作。

　　限于编者的学识水平与实践经验，书中不足之处在所难免，恳请读者和同行批评指正。

　　作者电子邮箱：zhuyilun2002@163.com

编　者

2015 年 2 月

目　录

Unit 1　A Brief Introduction of Electronic Technology (1)
  1.1　Text (1)
    1.1.1　History about Electronics (1)
    1.1.2　Introduce to Some Courses (2)
  1.2　Reading Materials (4)
    1.2.1　an Advertisement (4)
    1.2.2　Do You Know These Electronic Systems? (4)
    1.2.3　Domestic Power Plugs & Sockets (4)
    1.2.4　More Courses (5)
  1.3　Knowledge about Translation（翻译知识 1——科技英语的特点） (7)
  1.4　Exercises (8)
  1.5　课文译文 (9)
    1.5.1　电子技术历史 (9)
    1.5.2　一些课程介绍 (10)
  1.6　阅读材料参考译文 (11)
    1.6.1　招生广告 (11)
    1.6.2　你了解这些电子系统吗 (11)
    1.6.3　民用电源插座和插头 (11)
    1.6.4　更多课程介绍 (12)

Unit 2　Electrical Components (14)
  2.1　Text (14)
    2.1.1　Resistors (14)
    2.1.2　Capacitors (15)
    2.1.3　Inductors (16)
  2.2　Reading Materials (18)
    2.2.1　Nonlinear Resistors (18)
    2.2.2　Transformers (19)
    2.2.3　Various Low Voltage Apparatus (19)
  2.3　Knowledge about Translation（翻译知识 2——单词） (21)
  2.4　Exercises (22)
  2.5　课文译文 (23)
    2.5.1　电阻器 (23)
    2.5.2　电容器 (24)
    2.5.3　电感器 (25)
  2.6　阅读材料参考译文 (25)
    2.6.1　非线性电阻器 (25)

    2.6.2 变压器 ·· (25)
    2.6.3 各种低压电器设备 ·· (26)
Unit 3 Instruments ··· (28)
  3.1 Text ··· (28)
    3.1.1 Multimeters ·· (28)
    3.1.2 the Oscilloscope ··· (29)
  3.2 Reading Materials ··· (31)
    3.2.1 How Does an Oscilloscope Work ··· (31)
    3.2.2 Analog and Digital Oscilloscope ·· (32)
    3.2.3 Wattmeter ··· (32)
    3.2.4 Signal Generator ··· (33)
  3.3 Knowledge about Translation（翻译知识 3——非谓语动词 V-ing）··············· (34)
  3.4 Exercises ··· (35)
  3.5 课文参考译文 ··· (36)
    3.5.1 万用表 ·· (36)
    3.5.2 示波器 ·· (37)
  3.6 阅读材料参考译文 ··· (37)
    3.6.1 示波器是如何工作的 ·· (37)
    3.6.2 模拟示波器和数字示波器 ·· (38)
    3.6.3 瓦特表 ·· (38)
    3.6.4 信号发生器 ·· (39)
Unit 4 Electronic Components ··· (40)
  4.1 Text ··· (40)
    4.1.1 Semiconductor Diode ··· (40)
    4.1.2 NPN Bipolar Transistor ·· (41)
    4.1.3 MOS Transistors ··· (42)
    4.1.4 Ideal Operational Amplifier ·· (43)
  4.2 Reading Materials ··· (44)
    4.2.1 Audio Amplifiers ··· (44)
    4.2.2 the Transistor as a Switch ·· (45)
  4.3 Knowledge about Translation（翻译知识 4——非谓语动词 to V）················· (47)
  4.4 Exercises ··· (48)
  4.5 课文参考译文 ··· (49)
    4.5.1 半导体二极管 ·· (49)
    4.5.2 NPN 双极型晶体管 ··· (50)
    4.5.3 MOS 晶体管 ·· (50)
    4.5.4 理想运算放大器 ·· (51)
  4.6 阅读材料参考译文 ··· (51)
    4.6.1 音频放大电路 ·· (51)
    4.6.2 三极管用做开关 ·· (52)

Unit 5	Power Supplies	(53)
5.1	Text	(53)
	5.1.1　Information on Power Supplies	(53)
	5.1.2　Bridge (Full-wave) Rectifier	(53)
	5.1.3　Filter	(54)
	5.1.4　Zener Diode	(54)
	5.1.5　Linear Voltage Regulator	(55)
5.2	Reading Materials	(56)
	5.2.1　about the IEEE (Institute of Electrical and Electronics Engineers)	(56)
	5.2.2　Robots	(56)
	5.2.3　How Power Grids Work	(57)
5.3	Knowledge about Translation（翻译知识 5——非谓语动词 V-ed）	(60)
5.4	Exercises	(61)
5.5	课文参考译文	(62)
	5.5.1　关于（稳压）电源	(62)
	5.5.2　桥式（全波）整流器	(63)
	5.5.3　滤波器	(63)
	5.5.4　齐纳二极管（稳压管）	(63)
	5.5.5　线性稳压器	(63)
5.6	阅读材料参考译文	(64)
	5.6.1　关于 IEEE（电气电子工程师学会）	(64)
	5.6.2　机器人	(64)
	5.6.3　电力网是如何工作的	(65)
Unit 6	Linear Circuit Analysis	(67)
6.1	Text	(67)
	6.1.1　Electric Circuit	(67)
	6.1.2　Ohm's Law	(67)
	6.1.3　Kirchhoff's Law	(68)
	6.1.4　Circuit Analysis Techniques	(69)
	6.1.5　Sinusoidal Circuits	(69)
6.2	Reading Materials	(70)
	6.2.1　Circuit Breaker	(70)
	6.2.2　Information on Amplitude Modulation (AM)	(71)
	6.2.3　Thévenin's Theorem	(72)
	6.2.4　Apple iPod Classic 120 GB Silver	(73)
6.3	Knowledge about Translation（翻译知识 6——It 的用法）	(73)
6.4	Exercises	(75)
6.5	课文参考译文	(76)
	6.5.1　电路	(76)
	6.5.2　欧姆定律	(76)

· VII ·

- 6.5.3 基尔霍夫定律 (77)
- 6.5.4 电路分析方法 (77)
- 6.5.5 正弦电路 (78)
- 6.6 阅读材料参考译文 (78)
  - 6.6.1 电路断路器 (78)
  - 6.6.2 关于调幅 (79)
  - 6.6.3 戴维南定理 (79)
  - 6.6.4 苹果（公司）iPod classic 120 GB 银色 (79)

Unit 7 Integrated Circuit (81)
- 7.1 Text (81)
  - 7.1.1 Information on Integrated Circuits (81)
  - 7.1.2 Chip and Chip Holders (82)
  - 7.1.3 Bipolar Integrated Circuits & MOS Integrated Circuits (83)
  - 7.1.4 the Process of IC Design (83)
- 7.2 Reading Materials (84)
  - 7.2.1 Circuit Board (84)
  - 7.2.2 Circuit Delay (85)
  - 7.2.3 3G Phones to Use Sony FeliCa IC Chip (85)
  - 7.2.4 Electromagnetic Radiation and Lonosphere (86)
  - 7.2.5 Garmin Nuvi 260 3.5-Inch Portable GPS Navigator (87)
- 7.3 Knowledge about Translation（翻译知识 7——That 的用法）(88)
- 7.4 Exercises (89)
- 7.5 课文参考译文 (91)
  - 7.5.1 关于集成电路 (91)
  - 7.5.2 芯片和芯片插座 (91)
  - 7.5.3 双极型（晶体管）集成电路和 MOS 集成电路 (92)
  - 7.5.4 集成电路的设计过程 (92)
- 7.6 阅读材料参考译文 (92)
  - 7.6.1 电路板 (92)
  - 7.6.2 电路延迟 (93)
  - 7.6.3 用索尼公司 FeliCa 芯片的 3G 手机 (93)
  - 7.6.4 电磁辐射和电离层 (94)
  - 7.6.5 Garmin Nuvi 260 3.5 英寸便携式导航仪 (94)

Unit 8 Digital Logic Circuits (96)
- 8.1 Text (96)
  - 8.1.1 Number Systems (96)
  - 8.1.2 Logical Gates (96)
  - 8.1.3 the Flip-flops (97)
- 8.2 Reading Materials (99)
  - 8.2.1 74 Series Logic ICs (99)

	8.2.2	Registers	(100)
	8.2.3	Counter	(100)
	8.2.4	7-segment Display Drivers	(102)
8.3	Knowledge about Translation（翻译知识 8——Which 的用法）		(102)
8.4	Exercises		(104)
8.5	课文参考译文		(105)
	8.5.1	数字系统	(105)
	8.5.2	逻辑门	(105)
	8.5.3	触发器	(106)
8.6	阅读材料参考译文		(107)
	8.6.1	74 系列集成逻辑电路	(107)
	8.6.2	寄存器	(107)
	8.6.3	计数器	(108)
	8.6.4	7 段显示驱动（芯片）	(109)

Unit 9 Microcomputers (110)

9.1	Text		(110)
	9.1.1	Basic Computer	(110)
	9.1.2	the Motherboard	(110)
	9.1.3	the System Bus	(111)
	9.1.4	Main Memory	(112)
	9.1.5	BIOS (Basic Input/Output System)	(112)
9.2	Reading Materials		(114)
	9.2.1	Microcontroller	(114)
	9.2.2	about DNA Computers	(114)
	9.2.3	PLC	(115)
9.3	Knowledge about Translation（翻译知识 9——连词Ⅰ）		(116)
9.4	Exercises		(117)
9.5	课文参考译文		(119)
	9.5.1	基本型计算机	(119)
	9.5.2	主板	(119)
	9.5.3	系统总线	(119)
	9.5.4	主存（内存）	(120)
	9.5.5	BIOS（基本输入/输出系统）	(120)
9.6	阅读材料参考译文		(121)
	9.6.1	单片机（微控制器）	(121)
	9.6.2	DNA 计算机	(121)
	9.6.3	PLC（可编程逻辑控制器）	(122)

Unit 10 Programming the Computer (123)

10.1	Text		(123)
	10.1.1	C as a Structured Language	(123)

		10.1.2	MATLAB Language ··· (124)
		10.1.3	Assembly Language Instructions ··· (125)
		10.1.4	Introduce to operating systems ·· (126)
	10.2	Reading Materials ··· (128)
		10.2.1	What Does Operating System Do ·· (128)
		10.2.2	Introduction of Microsoft Certification Program (MCP) ············ (129)
		10.2.3	Object-Oriented Programming ·· (129)
	10.3	Knowledge about Translation （翻译知识 10——连词Ⅱ） ············· (130)
	10.4	Exercises ··· (131)
	10.5	课文参考译文 ··· (133)
		10.5.1	结构化语言 C ··· (133)
		10.5.2	MATLAB 语言 ··· (133)
		10.5.3	汇编语言指令 ··· (134)
		10.5.4	操作系统简介 ··· (134)
	10.6	阅读材料参考译文 ··· (135)
		10.6.1	操作系统做些什么 ··· (135)
		10.6.2	微软认证项目介绍 ··· (136)
		10.6.3	面向对象的编程 ··· (136)

Unit 11　Television ··· (137)

	11.1	Text ··· (137)
		11.1.1	about Television ··· (137)
		11.1.2	Color TV ··· (138)
		11.1.3	Getting the Signal to TV ·· (139)
	11.2	Reading Materials ··· (140)
		11.2.1	Digital TV ··· (140)
		11.2.2	LCD (Liquid Crystal Display) ··· (141)
		11.2.3	Pure Vision Plasma Display ·· (142)
	11.3	Knowledge about Translation（翻译知识 11——虚拟语气） ········· (142)
	11.4	Exercises ··· (144)
	11.5	课文参考译文 ··· (145)
		11.5.1	关于电视 ··· (145)
		11.5.2	彩色电视 ··· (146)
		11.5.3	电视机接收到的信号 ··· (146)
	11.6	阅读材料参考译文 ··· (147)
		11.6.1	数字电视 ··· (147)
		11.6.2	LCD（液晶显示器） ··· (147)
		11.6.3	纯平、等离子显示器 ··· (148)

Unit 12　Digital Camera ··· (149)

	12.1	Text ··· (149)
		12.1.1	Principle ··· (149)

· X ·

	12.1.2	Picture Quality	(149)
	12.1.3	Features	(150)
	12.1.4	Memory and Connectivity	(151)
12.2	Reading Materials		(153)
	12.2.1	Digital Camcorders	(153)
	12.2.2	Video Compression	(153)
	12.2.3	X3 Technology	(154)
12.3	Knowledge about Translation（翻译知识 12——倒装）		(155)
12.4	Exercises		(156)
12.5	课文参考译文		(157)
	12.5.1	原理	(157)
	12.5.2	图像质量	(158)
	12.5.3	特点	(158)
	12.5.4	存储器和连接	(158)
12.6	阅读材料参考译文		(159)
	12.6.1	数码摄像机	(159)
	12.6.2	视频压缩	(159)
	12.6.3	X3 技术	(160)

Unit 13 Internet-based Communication (161)

13.1	Text		(161)
	13.1.1	Instant Messaging	(161)
	13.1.2	Internet Telephony & VoIP（Voice over Internet Protocol）	(162)
	13.1.3	E-mail	(162)
	13.1.4	Videoconference	(163)
	13.1.5	SMS & Wireless Communications	(163)
13.2	Reading Materials		(164)
	13.2.1	Server	(164)
	13.2.2	about Web Page	(165)
	13.2.3	the Cell Approach	(166)
	13.2.4	Inside a Cell Phone	(166)
	13.2.5	Nokia N95	(167)
13.3	Knowledge about Translation（翻译知识 13——否定形式）		(168)
13.4	Exercises		(169)
13.5	课文参考译文		(171)
	13.5.1	即时消息（网上聊天）	(171)
	13.5.2	网上通话和网络电话	(171)
	13.5.3	电子邮件	(172)
	13.5.4	视频会议	(172)
	13.5.5	短消息服务和无线通信	(172)
13.6	阅读材料参考译文		(173)

		13.6.1 服务器	(173)
		13.6.2 网页	(173)
		13.6.3 手机（蜂窝）技术	(174)
		13.6.4 手机的内部	(174)
		13.6.5 诺基亚 N95	(175)

Unit 14 Electrical Appliances .... (176)
  14.1 Text .... (176)
    14.1.1 the Refrigerator .... (176)
    14.1.2 the Air Conditioner .... (177)
    14.1.3 the Microwave Oven .... (177)
  14.2 Reading Materials .... (179)
    14.2.1 Gas and Propane Refrigerators .... (179)
    14.2.2 History about Microwave Oven .... (180)
    14.2.3 Vacuum Cleaner .... (180)
    14.2.4 Embedded System .... (182)
  14.3 Knowledge about Translation（翻译知识14——分离现象）..... (182)
  14.4 Exercises .... (183)
  14.5 课文参考译文 .... (184)
    14.5.1 冰箱 .... (184)
    14.5.2 空调 .... (185)
    14.5.3 微波炉 .... (185)
  14.6 阅读材料译文 .... (186)
    14.6.1 汽油和丙烷冰箱 .... (186)
    14.6.2 微波炉的由来 .... (186)
    14.6.3 真空吸尘器 .... (187)
    14.6.4 嵌入式系统 .... (188)

Unit 15 I/O Devices .... (189)
  15.1 Text .... (189)
    15.1.1 Keyboards .... (189)
    15.1.2 Mice (or Mouse) .... (190)
    15.1.3 Inkjet Printers .... (191)
    15.1.4 the Laser Printer .... (192)
  15.2 Reading Materials .... (194)
    15.2.1 Touchscreen .... (194)
    15.2.2 Colour Lasers .... (195)
    15.2.3 Electronic Circuit Simulation .... (196)
  15.3 Knowledge about Translation（翻译知识15——省略和插入语）..... (197)
  15.4 Exercises .... (199)
  15.5 课文参考译文 .... (200)
    15.5.1 键盘 .... (200)

		15.5.2	鼠标	(201)
		15.5.3	喷墨打印机	(202)
		15.5.4	激光打印机	(202)
	15.6	阅读材料参考译文		(203)
		15.6.1	触摸屏	(203)
		15.6.2	彩色激光打印机	(204)
		15.6.3	电子电路仿真	(205)
Unit 16	Multimedia Technology			(206)
	16.1	Text		(206)
		16.1.1	What is Multimedia	(206)
		16.1.2	Multimedia Assets	(206)
		16.1.3	Multimedia Applications	(207)
	16.2	Reading Materials		(209)
		16.2.1	DVD	(209)
		16.2.2	News: Vivid Animations Help Students with Science	(211)
		16.2.3	about Graphics Cards	(211)
		16.2.4	DirectX	(212)
	16.3	Knowledge on Writing a Research Paper（科技论文写作知识）		(213)
	16.4	Exercises		(215)
	16.5	课文参考译文		(217)
		16.5.1	什么是多媒体	(217)
		16.5.2	多媒体资源	(217)
		16.5.3	多媒体的应用	(218)
	16.6	阅读材料参考译文		(219)
		16.6.1	DVD	(219)
		16.6.2	新闻：学生通过生动的仿真（实验）学科学	(220)
		16.6.3	关于图像卡（或译做视频卡）	(220)
		16.6.4	DirectX	(221)
Unit 17	User's Manual			(222)
	17.1	Text		(222)
		17.1.1	Introduction to MiraScan	(222)
		17.1.2	Scanning Reflective Originals	(222)
		17.1.3	Understanding MiraScan Functions	(223)
	17.2	Reading Materials		(226)
		17.2.1	NE555	(226)
		17.2.2	AD574A	(227)
		17.2.3	Cover letter	(228)
		17.2.4	Resume	(230)
	17.3	Knowledge on Writing User's Manual（用户说明书写作知识）		(231)
	17.4	Exercises		(232)

17.5 课文参考译文 ··· (233)
    17.5.1 MiraScan 介绍 ··· (233)
    17.5.2 扫描不透明的原件 ·· (233)
    17.5.3 MiraScan 功能 ··· (234)
17.6 阅读材料参考译文 ·· (235)
    17.6.1 NE555 ··· (235)
    17.6.2 AD574A ·· (236)
    17.6.3 求职信 ·· (236)
    17.6.4 简历 ··· (237)

Appendix A Reference Answers ··· (239)
Appendix B Technical Vocabulary Index ·· (259)
Reference ··· (270)

Unit 1 A Brief Introduction of Electronic Technology

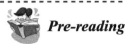
Pre-reading

Read the following passage, paying attention to the question.
1) Who invented the radio?
2) When was the real electronics started?
3) What courses should a student majoring in the electronics study?
4) What career will a student majoring in the electronic technology follow?

1.1 Text

1.1.1 History about Electronics

There can be no doubt that the 1900s is remembered as the electronic century. Of course there have been other great advances, in medicine, in transport, in science, in commerce, and many other fields, but where would they have been without the instruments and devices that electronics has provided? How would you see a 3D virtual reality image of your beating heart with no electronics? How would you get money out of the bank on a Sunday night without electronics? Would you go to a pop concert that had no amplifiers, large screens or lighting effects? Don't say you would rather watch TV – there would not be any.

Electronics in the early 20th century started thriving at a greater speed unlike the pre-20th century developments. The radio invented by the Italian genius Marconi and the work of Henry Hertz opened the road to further discoveries and inventions. In the first decade the new thing that was welcomed to the technical world was the vacuum tube. The vacuum tubes at that time worked as a miraculous component for the radio devices.

The invention of the television was a miraculous thing for the mankind. It was revolution in both communication technology and also for the world media. The distances between the continents did not seem to be far enough. The credit goes to the British engineer John Logic Baird who followed the footprints of Marconi and tried to send the images in the same way as the speech. After a long experiment he found that a series of static pictures if sent within a small interval of time in between them, seem to be moving.

The real electronics what it is called today was actually started after the discovery of the transistor effect. Transistor opened the road for the electronics and more importantly it opened the road for the computing world. Computers of various types started hitting the market and the research works got a boost.

Some other problems were also there like the assembling of the electronic components on a single mother board. Jack Kilby in Texas Instruments found a very nice solution. He suggested to throw away all the wires and tried to connect the resistors, capacitors and transistors on the same piece of wafer internally. Surprisingly his ideas worked and gave birth to the integrated circuit industries.

1.1.2 Introduce to Some Courses

As a student majoring in the electronic technology, you will study many courses such as:

1. Direct Current Circuits & Alternating Current circuits

This course covers the fundamental theory of passive devices (resistor, capacitor and inductor) and electrical networks supplied by a DC source, and then an introduction to the effects of alternating voltage and current in passive electrical circuits is given. This module also covers DC machines, three phase machines and transformers.

2. Analog Electronics

This module introduces the characteristics of semiconductor devices in a range of linear applications and electronic circuits consisting of these devices (Fig 1.1). The following specific topics are covered. Semiconductor diodes: PN junction diodes, special purpose diodes; Transistors: field effect and bipolar transistors; Signal amplifiers: practical amplifiers, biasing circuits, operational amplifiers circuit; Other circuits: rectification, regulation and DC power supplies.

Fig 1.1 electronic circuit

3. Digital Electronics

In this unit the following topics are covered: basic concepts about Logic circuits, number representations, combinatorial logic circuits, sequential logic circuits, introduction to CMOS digital circuits, logic operations theorems and Boolean algebra, number operations (binary, hex and integers), combinatorial logic analysis and synthesis, sequential logic analysis and synthesis, registers, counters, bus systems, CAD tools for logic design.

4. Microcontroller Systems

The use of computers and microcontrollers is now found in every field of the electronics industry. This use will continue to grow at a rapid pace as computers become more complex and powerful. The ability to program these devices will make a student an invaluable asset to the growing electronic industry. This module enables the student to program a simple microcontroller to perform typical industrial tasks. Assembler and C are used to program the MPU (Microprocessor Unit). The student will set up the internal devices such as RS232 port, timer, interrupts, counters, I/O ports, ADC etc. The program will then use these devices for control operations.

5. Computer Programming for Engineering Applications

It is a continuation of more advanced programming techniques. The language of C will be used for teaching purposes. Emphasis is towards the use of programming for engineering applications and problem solving.

The electronic technology will provide a sound educational foundation to enable graduates to follow a career in: electrical engineering; power and control engineering; electronics; computer engineering; telecommunications engineering etc.

Technical Words and Phrases

amplifier	[ˈæmplɪfaɪ(ə)]	n. [电工]扩音器，放大器
analog	[ˈænəlɔg]	n. 类似物，相似体，（计算机）模拟
career	[kəˈrɪə(r)]	n. 事业，生涯
characteristics	[kærɪktəˈrɪstɪk]	adj. 特有的 n. 特性，特征，特征值
circuit	[ˈsəːkɪt]	n. 电路 vt. 接成电路（绕……环行）
image	[ˈɪmɪdʒ]	n. 图像，肖像，映像 vt. 想象，作……的像
instrument	[ˈɪnstrumənt]	n. 工具，仪器，乐器
interrupt	[ɪntəˈrʌpt]	vt. 打断，中断 vi. 打断 n. （发给电脑的）中断信号，中断
medicine	[ˈmeds(ə)n]	n. 药，医学，内科学，内服药 vt. 给……用药
miraculous	[miˈrækjuləs]	adj. 奇迹的，不可思议的
module	[ˈmɔdjuːl]	n. 模数，模块，登月舱，指令舱，此处指课程模块
passive	[ˈpæsɪv]	adj. 被动的，此处指无源的
program	[ˈprəugræm]	n. 节目，程序 vi. 安排节目，编程序
semiconductor	[semɪkənˈdʌkt]	n. [物]半导体
static	[ˈstætɪk]	adj. 静态的，静力的
technique	[tækˈniːk]	n. 技术，技巧，方法，表演法，手法，工艺，技艺
thrive	[ˈθraɪv]	v. 兴旺，繁荣，茁壮成长，旺盛
transistor	[trænˈsɪstə(r)]	n. [电子]晶体管；晶体管（半导体管，晶体管收音机）
virtual	[ˈvɜːtjuəl]	adj. 虚的，实质的，虚拟的，[物]有效的，事实上的

AC(alternating current)	交流电（流）
Applications	应用，应用软件
DC(direct current)	直流电（流）
electronic technology	电子技术
I/O port	输入/输出端口
major in	（在大学里）主修
pop concert	流行音乐会
power supplies	电源
the credit goes to…	归功于……

Notes to the Text

1. MPU   Microprocessor Unit 微处理器的缩写
2. microcontroller   微控制器，微处理器，单片机（有时用 microprocessor）

1.2 Reading Materials

1.2.1 an Advertisement

The new bachelor of engineering qualification has been designed with the needs of students and industry firmly in mind. It utilizes practical engineering examples and projects to enable students to place their knowledge in context. The degree includes the study of commercial, managerial and ethical topics as requested by the employers of professional engineers.

The faculty of Science and Engineering has a close relationship with a broad range of engineering companies. This relationship is invaluable in ensuring that the Bachelor of Engineering programmed is relevant to the present and future needs of engineering employers.

Facilities available include six computer labs including one of Australia's largest CAD/CAM Suites, electrical, electronic, telecommunications, instrumentation, mechanics, thermodynamics and pneumatics laboratories. Also available for student is a world class precision machine shop which has a wide range of up-to-date machine tools.

（这是一则国外大学的招生广告的摘录，所以它的用词十分简洁、生动。）

1.2.2 Do You Know These Electronic Systems?

Some electronic systems are familiar from everyday life. For example, we encounter radios, televisions, telephones, and computers on a daily basis. Other electronic systems are present in daily life, but are less obvious. Electronic systems control fuel mixture and ignition timing to maximize performance and minimize undesirable emissions from automobile engines. Electronics in weather satellites (Fig 1.2) provide us with a continuous detailed picture of our planet.

Fig 1.2   weather satellite

Still other systems are even less familiar. For example, a system of satellites known as the Global Positioning System (GPS) has been developed to provide three-dimensional information for ships, aircrafts and cars anywhere on earth. This is possible because signals emitted by several satellites can be received by the vehicle，by comparing the time of arrival of the signals and by using certain information contained in the received signals concerning the orbits of the satellites, the position of the vehicle can be determined.

Other electronic systems include the air-traffic control system, various radars, compact-disc (CD) recording equipment and players, manufacturing control systems, and navigation systems.

1.2.3 Domestic Power Plugs & Sockets

In most countries, household power is single-phase electric power, in which a single phase conductor brings alternating current into a house, and a neutral returns it to the power supply.

Domestic power plugs and sockets are devices that connect the home appliances and portable light fixtures commonly used in homes to the commercial power supply so that electric power can flow to them. Many plugs and sockets include a third contact used for a protective earth ground, which only carries current in case of a fault in the connected equipment.

Power plugs are male electrical connectors that fit into female electrical sockets. They have contacts that are pins or blades that connect mechanically and electrically to holes or slots in the socket. Plugs usually have a phase or hot or live contact, a neutral contact, and an optional earth or Ground contact. Many plugs make no distinction between the live and neutral contacts, and in some cases they have two live contacts. The contacts may be steel or brass, either zinc, tin or nickel plated.

Power sockets are female electrical connectors that have slots or holes which accept the pins or blades of power plugs inserted into them and deliver electricity to the plugs. Sockets are usually designed to reject any plug which is not built to the same electrical standard. Some sockets have one or more holes that connect to pins on the plug.

The domestic power variolls plugs and sockets used in some countries are shown in Fig 1.3.

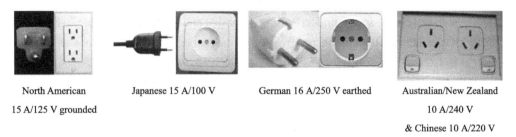

North American 15 A/125 V grounded   Japanese 15 A/100 V   German 16 A/250 V earthed   Australian/New Zealand 10 A/240 V & Chinese 10 A/220 V

Fig 1.3   domestic power Various Plugs & Sockets

1.2.4 More Courses

1. Signals and Systems

This unit aims to teach some of the basic properties of many engineering signals and systems and the necessary mathematical tools that aid in this process. The particular emphasis is on the time and frequency domain modeling of linear time invariant systems. The concepts learnt in this unit will be heavily used in many units of study (in later years) in the areas of communication, control, power systems and signal processing. A basic knowledge of differentiation and integration, differential equations, and linear algebra is assumed.

The following topics are covered. Continuous-time signals: classification and properties; Basic properties of systems: linearity, time-invariance, causality, and stability. Linear time-invariant (LTI) systems: characterization by differential equations (including state space formulation), and the convolution integral. Fourier series and Fourier Transform: definition, properties, frequency response and analysis of LTI systems based on Fourier transform, sampling, correlation and power spectral density; Laplace transform: definition, properties, and analysis of LTI systems based on Laplace transform, solution of state space equations using Laplace transform.

2. Circuit Theory and Design

This unit of study assumes a basic knowledge of elementary circuit theory and operational amplifiers provided by earlier units. One aim of the unit is to enhance understanding of key aspects of the theory of electric circuits. The main goal, however, is to equip students with the specialist knowledge to design active analog filters, to have an understanding of passive network design and to be in a good position to undertake further self study as required.

The specific topics covered include the following: Fundamental concepts in circuit theory: network functions, characteristic frequencies; Types of filter: lowpass, bandpass etc. Review of operational amplifiers: design of first and second order filters using operational amplifiers. cascade design. Filter characteristics: Butterworth, Chebyshev, frequency transformations in design, sensitivity design of passive LC ladder filters, a brief introduction to switched capacitor filters.

3. Control Theory

This unit is concerned with the application of feedback control to continuous-time, linear time-invariant systems. The emphasis is on fundamental theory rather than applications. Some background in linear systems theory and the Laplace transform is assumed. The prime aim of this unit of study is to develop a sound understanding of basics and a capacity for research and inquiry. Completion of the unit will facilitate progression to advanced study in the area and to work in industrial control.

The following topics are covered. History of control: Modeling of physical processes, state variables and differential equations, dynamic response, review of Laplace transform, transfer functions and block diagrams, poles and zeroes; Design specifications in the time domain: basic feedback principles, effect of feedback on sensitivity and disturbance rejection, steady state accuracy and stability, the Routh criterion, proportional integral and derivative control; Design using the root locus: rules for sketching root locus, lead and lag compensators, analogue and digital implementation of controllers; Frequency response: the Nyquist stability criterion, gain and phase margins, compensator design in the frequency domain; An introduction to state space design for single-input single-output systems: eigenvalues, zeroes and transfer functions, state variable feedback and design of estimators.

4. Communications Electronics and Photonics

This unit of study provides an introduction to the modeling and design of transmitters and receivers for electronic and optical communication subsystems. Students are expected to have a grasp of basic concepts related to electronics and circuits.

The following topics are covered: Electronic oscillators: RC, LC, crystal oscillators, tuned electronic amplifiers, frequency selectivity, feedback amplifiers; Electronic modulation and demodulation circuits: amplitude, frequency and phase modulation and demodulation, phase locked loops; Electronic mixers: high frequency, RF and microwave communication amplifiers; Photonic devices and models: semiconductor optical properties, semiconductor lasers and light emitting diodes, laser modes, output spectra, single-mode selection, distributed feedback lasers; Electro-optic modulation of light: optical amplifiers, photo detectors, avalanche photodiodes, optical receiver front-end circuit design, basic opto-electronic link.

5. Power Electronics and Drives

This unit of study is concerned with the operating principles of DC machines and DC power control techniques with particular reference to DC machine drives. A background in basic electrical and magnetic circuit theory is assumed. Completion of this unit will facilitate progression to advanced study or work in electrical power engineering.

The following topics are covered. electrical characteristics of separately excited, series, shunt and compound generators, voltage control of generators, electrical characteristics of separately excited, series, shunt and compound motors, starting and speed control of DC motors, static switches, diode rectifiers, AC-DC converters, DC-DC switching converters, Buck, Boost and Buck-Boost converters, flyback converters.

1.3 Knowledge about Translation
（翻译知识 1——科技英语的特点）

科技英语具有下列 4 个特点。

1. 复杂长句多

科技文章要求叙述准确，用词严谨，因此一句话里常常包含多个分句，这种复杂且长的句子居科技英语难点之首，阅读翻译时要按汉语习惯来加以分析，以短代长，化难为易。

Of course there have been other great advances in medicine, in transport, in science, in commerce and many other fields, but where would **they** have been without the instruments and devices that electronics has provided? 当然在许多其他领域如医药、交通、科学、商业等行业也取得很大进步，但如果没有电子提供的仪器和器件，这些行业能取得这么大的进步吗？

这是一个带有一对并列句和一个从句组成的长句。其中 they 指的都是 great advances。在科技英语中，要注意 it, that, which 等词的指代，有时要结合自己的专业知识来翻译。

2. 被动语态多

科技英语为了强调所论述的客观事物，常把它放在句子的首位，以突出其重要性。

The language of C **will be used** for teaching purposes. Emphasis is towards the use of programming for engineering applications and solving problem.采用 C 语言进行教学时，重点放在如何运用编程技术解决工程应用的实际问题。

在翻译中可以根据中文的习惯，不一定要译出被动语态。

The following topics are covered. 主要包括以下内容。

3. 非谓语动词多

英语每个简单句中，只能用一个谓语动词，如果有几个动词就必须选出主要动词当谓语，而将其余动作用非谓语动词形式（V-ing, V-ed ,to V 三种形式）表示，才能符合英语语法的要求。

This module enables the student **to program** a simple microcontroller **to perform** typical industrial tasks. 这个模块（教学）使学生能对一个简单的微处理器进行编程（使其）执行典型的工业任务。

这里 to program, to perform 都是非谓语动词形式描述动作。

非谓语动词也常用做定语等。

The ability **to program these devices** will make a student an invaluable asset to the **growing** electronic industry. 对日益增长的电子工业来说，一个具有微处理器编程能力的学生将会是无价的人才。

这里 to program theses devices 作定语，修饰前面的 ability; growing 作定语，修饰后面的 electronic industry。这里 asset 原意为资产、有用的东西，我们可根据上下文译成人才。

4．词性转换多

英语单词有不少是多性词，即既是名词，又可作为动词、形容词、介词或副词，字形无殊，功能各异，阅读时也很容易造成曲解。例如：light。

用做名词：	high light	强光，精华
	safety light	安全灯
用做形容词：	light industry	轻工业
	light room	明亮的房间
	light blue	淡蓝色
	light coating	薄涂层
用做动词：	light up the lamp	点灯
用做副词：	travel light	轻装旅行

因此，在翻译时要根据上下文的意思选取词意。

1.4　Exercises

1. Put the Phrases into English

① 直流电路　　　　　　⑥ 微处理器
② 放大器（扩音器）　　⑦ 电气工程
③ 模拟电子技术　　　　⑧ 能源工程（或电力工程）
④ 半导体二极管　　　　⑨ 通信工程
⑤ 晶体管效应　　　　　⑩ 内部器件

2. Put the Phrases into Chinese

① assembler language　　　　⑥ logic gates
② alternating current circuits　⑦ 3D virtual reality image
③ passive electrical circuits　　⑧ computer programming
④ three phase circuits　　　　⑨ major in
⑤ digital eletronics　　　　　⑩ advanced programming techniques

3. Sentence Translation

① Would you go to a pop concert that had no amplifiers, large screens or lighting effects?

② The credit goes to the British engineer John Logic Baird who followed the foot prints of Marconi and tried to send the images in the same way as the speech.

③ The real electronics what it is called today was actually started after the discovery of the transistor effect.

④ Surprisingly his ideas worked and gave birth to the integrated circuit industries.

⑤ This module introduces the characteristics of semiconductor devices in a range of linear applications.

⑥ The use of computers and microcontrollers is now found in every field of the electronics industry.

⑦ This module enables the student to program a simple microcontroller to perform typical industrial tasks.

⑧ The program will then use these devices for control operations.

⑨ Emphasis is towards the use of programming for engineering applications and problem solving.

⑩ The electronic technology will provide a sound educational foundation to enable graduates to follow a career in electrical engineering.

4. Read and translate it into Chinese

① The study of electric circuits is fundamental in electrical engineering education, and can be quite valuable in other disciplines as well. The skills acquired not only are useful in such electrical engineering areas as electronics, communications, microwaves, control, and power systems but also can be employed in other seemingly different fields.

② The impact of digital integrated circuits on our modern society has been pervasive. Without them, the revolution of current computer and information-technology would not exist. Digital integrated circuits represent the most important enabling technology in this revolution. This is largely true because of the immense amount of signal and computer processing which can be realized in a single integrated circuit.

1.5 课 文 译 文

1.5.1 电子技术历史

毫无疑问，20世纪是电子技术的世纪。当然许多其他行业如医药、交通、科学、商业等也取得很大进步，但如果没有电子技术所提供的仪器和设备，这些行业能取得这么大的进步吗？没有电子技术，你无法看到自己正在跳动的心脏的逼真的三维虚拟图像；没有电子技术，你就无法在星期日的晚上从银行取钱。你愿意去参加一个没有音响放大器、没有大屏幕或灯光效果的流行音乐会吗？不要说你宁愿在家看电视——没有电子技术也就没有电视。

与20世纪前不同，在20世纪早期，电子技术开始有了较快的发展。首先意大利天才马可尼（Marconi）发明的无线电和亨利·赫兹（Henry.Hertz）的工作为电子技术进一步的发明创造开辟了道路。在20世纪第一个十年中最受技术世界欢迎的新东西是真空管，在那时真空管是无线电设备中一个奇妙的器件。

对于人类来说，电视的发明也是一个奇迹。电视带来了通信技术和世界传媒的革命。有了电视，洲与洲的距离似乎不再遥远。电视的发明应归功于英国工程师约翰·罗杰克·贝尔德（John Logic Baird），他追随马可尼（Marconi）的足迹，想用与传送声音相同的方式传送图像。经过长时间实验后，他发现如果以很短的时间间隔发送一组静态的图片，看起来就像是活动的图像。

今天所说的电子技术实际上是在发现晶体管效应以后开始（发展）的。晶体管为电子技术开辟了道路，更重要的是它为计算机世界开辟了道路。各种类型的计算机开始在市场上出现，研究工作进入一个迅速发展的时代。

在电子技术发展过程中还存在其他的问题，如电子器件在一块主板上的安装问题。对此德克萨斯仪器公司的杰克·柯比（Jack.Kilby）找到了很好的答案。他提议不用任何导线，把电阻、电容和晶体管在同一片晶片内部连接起来，令人不可思议的是他的想法成功了，从此诞生了集成电路工业。

1.5.2 一些课程介绍

作为一个电子技术专业的学生，要学习下列课程。

1．直流电路与交流电路

这门课程包括无源元器件（电阻、电容和电感）的基本理论和用直流电源供电的电路网络，接着介绍无源电路中的交流电流和交流电压的作用，这个课程模块还包括直流电机、三相电机和变压器。

2．模拟电子技术

这个课程模块介绍半导体器件在线性应用范围中的特征和由这些器件组成的电路（如图 1.1 所示），内容包括半导体二极管：PN 结二极管、特殊二极管；三极管：场效应三极管、晶体三极管；信号放大电路：实际放大电路、偏置电路、运算放大器电路；其他电路：整流、稳压、直流电压源电路。

3．数字电子技术

这个单元学习以下的内容：逻辑电路的基本概念、数字表示方法、组合逻辑电路、时序逻辑电路、CMOS 数字电路、逻辑运算定律和布尔代数、数字运算（二进制、十六进制、整数）组合逻辑电路的分析与综合、时序逻辑电路的分析与综合、寄存器、计数器、总线系统以及逻辑电路设计中的计算机辅助设计工具（软件）。

4．微处理器系统

当前，计算机及微处理器在电子工业的各个领域中应用十分广泛，随着计算机变得更加复杂和功能强大，微处理器的应用将继续快速增长。对日益增长的电子工业来说，一个具有微处理器编程能力的学生将会是无价的人才。这个模块中安排学生对一个简单的微处理器进行编程来完成工业上典型的控制任务。用汇编语言和 C 语言对微处理器进行编程时，学生将用到一些内部的器件如 RS232 接口、定时器、中断器件、计数器、输入/输出口、模/数转换器等，将利用这些器件通过编程完成控制（系统）等操作。

5．计算机编程及其在工程中的应用

该课程继续学习更高级的编程技术，教学中采用 C 语言，重点放在如何运用编程技术解决工程应用的实际问题。

电子技术专业将为毕业生打下一个牢固的基础，学生毕业后可以从事的行业有：电气工程、电力能源和控制工程、电子技术、计算机工程、通信工程等。

1.6 阅读材料参考译文

1.6.1 招生广告

考虑到学生和工业上的需要，（我们）开设了一种新的工程学士（学位）专业。该专业利用工程实际例子和项目使学生掌握实用的知识，专业教学内容还包括专业工程师所需要的商业、管理和道德教育。

科学工程系的全体教职员与很多行业的工程公司保持着紧密的联系，这种联系保证了（我们的）工程学士学位课程安排是符合现在和未来的工程雇主（对雇佣人的）要求的（即有很好的就业前景）。

我们拥有6个计算机实验室，其中包括属于澳大利亚最大的CAD/CAM配套中心之一的实验室，以及电气、电子、通信、仪器、机械、热力学和气体力学实验室。另外还有一个配有很多最新机械仪器的世界级精密仪器车间也对学生开放。

1.6.2 你了解这些电子系统吗

有些电子系统在日常生活中很常见，例如收音机、电视机、电话、家用计算机。有些电子系统也是日常生活中常用的，但很少引起人们注意，如汽车中用电子系统来控制燃料混合和点火时间，使发动机的性能可以达到最佳，尾气排放最少；又如人造气象卫星（如图1.2所示）中的电子系统为我们提供地球（周围气象）的连续、详细的图像。

另有一些电子系统可能更少见，例如称为全球定位系统（GPS）的卫星系统，用于为位于地球上任何位置的船舶、飞机和汽车提供三维定位信息，当它们接收到来自几颗卫星所发射的信号后，通过比较信号到达的时间和信号中所含的卫星轨道的信息，可以确定它们自己的位置。

其他电子系统还有航空飞行控制系统、各种雷达、光盘录音设备和播放器、制造业生产控制系统和导航系统。

1.6.3 民用电源插座和插头

很多国家，民用电源是单相电源，用一根火线（相线）和一根中线将交流电送入民居。

民用电源插头和插座是用来连接家用电器和家中所用的可移动的照明设施的，使家电和照明设施可以有电源供电，有电流流过。许多插头和插座还有用来保护接地的第三个接触头（孔）。保护接地是当所接的电器设备出现（漏电）故障时引导电流流入地的。

电源插头是电气接触头（俗称公插头），可以插入电源插座（俗称母插座）。圆形或扁平形的插头插入插孔或插槽中，使它们相互接触。插头通常有相线（或火线）、中线和接地的插头。很多插头并不区分相线和中线的插头，有的插头有两个相线插头。插头可以是用钢、青铜、锌、锡或镍材料制成的。

插座是另一种电气接触器，插座上有插槽或插孔让电源插头上的圆形或扁平形的插头插入，把电源中的电能输出给插头。插座通常设计成不允许不是相同电气标准的插头插入。有些插座有一组或多组插孔。

图1.3给出部分国家所用的民用电源插座和插头的照片。

1.6.4 更多课程介绍

1. 信号与系统

这个单元讲解许多工程信号和系统的基本性质以及在信号和系统处理中必需的数学工具，重点放在线性时不变系统的时域和频域模型上。在这个单元中所学的概念将在以后学习的通信、控制、电力系统和信号处理等领域的许多单元中用到，学习这个单元需要具有微分、积分、微分方程和线性代数等基础知识。

主要内容包括连续时间信号：分类及性质；系统的基本性质：线性、时不变性、因果性和稳定性；线性时不变系统：由微分方程（包括状态方程）描述的特征和卷积；傅里叶级数和傅里叶变换：定义、性质、频率响应和基于傅里叶变换的线性时不变系统的分析、采样、相关性和功率谱密度；拉普拉斯变换：定义、性质和基于拉普拉斯变换的线性时不变系统的分析、用拉普拉斯变换求解状态方程。

2. 电路理论和设计

这个单元学习之前必须具备由前期课程所提供的基本电路理论和运算放大器知识。本单元的学习目标是增强对电路理论的主要方面的理解。而主要目的是：使学生掌握专业知识，可以从事有源模拟滤波器的设计，理解无源网络设计方法，为今后进一步自学打下良好的基础。

主要包括的内容有电路理论的基本概念：网络函数、特征频率；滤波器类型：低通、带通滤波器等；运算放大器的讨论：用运算放大器设计的一级、二级滤波器、电路串联（级联）设计；几种典型的滤波器：Butterworth（巴特沃斯）、Chebyshev（契比雪夫）滤波器、设计中的频率变换、无源 LC 梯形滤波器的灵敏度设计，并对开关电容滤波器做简短的介绍。

3. 控制理论

这个单元是讲授关于连续、线性时不变系统的反馈控制的应用，重点是基本的理论而不是应用。这个单元的学习要求学生具有线性系统理论和拉普拉斯变换的基础。这个单元学习的主要目的是（使学生）在基本理论和进一步研究的能力方面打下一个坚实的基础，这个单元的学习将促进学生在本领域的进一步学习和今后在工业控制行业的工作。

主要内容包括控制理论的历史；物理过程的模型化方法：状态变量和微分方程、动态响应、拉普拉斯变换的讨论、传递函数和方框图、极点和零点；时域系统的设计方法：基本反馈原理，反馈对灵敏度、抗干扰性、稳态精度和稳定性的影响、Routh（罗斯）判断准则；比例、积分和微分控制；用根轨迹法设计：根轨迹作图规则、超前和滞后补偿、模拟控制器和数字控制器的实现；频率响应：Nyquist（纳奎斯特）稳定性判据、增益裕度和相位裕度、频域的补偿设计；并介绍了单一输入/输出系统的状态方程设计方法：本征值、零点和传递函数、状态变量反馈和计算方法的设计。

4. 通信电子学和光学

本单元介绍电子和光学通信子系统的发射机和接收机的建模和设计方法，（本单元的学习）要求学生已掌握有关电子和电路的基本概念。

主要包括的内容有：电子振荡器：RC、LC、晶体振荡器、调谐电子放大器、频率选择、反馈放大器；电子调制和解调电路：幅度、频率和相位调制和解调、锁相环；电子混合器：

高频、射频和微波通信放大器；光学器件和模型：半导体光学性质、半导体激光和发光二极管、激光模态、输出光谱、单一模态选择、分布式反馈激光器；光的电子-光学调制：光学放大器、光电探测器、雪崩光敏二极管、光学接收器的前置电路设计、基本的光-电子连接。

5．功率电子学和驱动

这个单元的学习涉及直流电机的工作原理和与直流电机驱动相关的直流功率控制技术，要求学生已学过基本的电磁场电路理论。这个单元的学习将促进学生在电力行业中的进一步学习和工作。

主要内容包括：他励、串励、并励和复励式直流发电机的电气特性、直流发电机的电压控制，分励、串励、并励和复励直流电动机的电气特性、直流电动机的启动和速度控制，静态变换，二极管整流，交流-直流转换，直流-直流变换器，Buck（巴克）、Boost（巴斯特）和巴克-巴斯特转换器，逆变转换器。

Unit 2 Electrical Components

 Pre-reading

Read the following passage, paying attention to the question.
1) What is a resistor?
2) What is Ohm law?
3) What is a capacitor and what is a inductor?

2.1 Text

2.1.1 Resistors

A resistor is an electrical component that resists the flow of electrical current. The amount of current (I) flowing in a circuit is directly proportional to the voltage across it and inversely proportional to the resistance of the circuit. This is Ohm law and can be expressed as a formula: $I = \dfrac{U_R}{R}$. The resistor is generally a linear device and its characteristics form a straight line when plotted on a graph.

Resistors are used to limit current flowing to a device, thereby preventing it from burning out, as voltage dividers to reduce voltage for other circuits, as transistor biasing circuits, and to serve as circuit loads.

Generally, resistors (Fig 2.1) consist of carbon composition, wire-wound, and metal film. The size of resistors depends on power ratings. Larger sizes are referred to as power resistors. Variable resistors are adjustable: rheostats, potentiometers, and trimmer pots. Precision resistors have a tolerance of 1% or less.

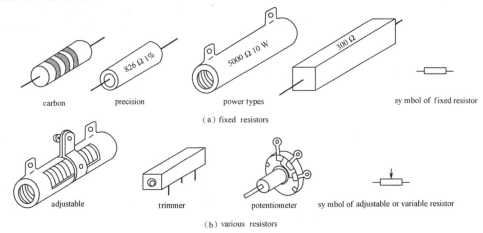

Fig 2.1 fixed and variable resistors

If you are a bit serious about the electronics hobby I recommend learning the "Color Code". It makes a lot easier. The same color code is used for everything else, like coils, capacitors etc. Again, just the color code associated with a number, like: black=0 brown=1 red=2, etc.

Fig 2.2 is an example. It is a 4-band resistor. The first band is the tens values; the second band gives the units; the third band is a multiplying factor, the factor being 10's band value. The fourth band gives the tolerance of the resistor. No band implies a tolerance of ± 20%, a silver band means the resistor has a tolerance of ± 10% and a gold band has the closest tolerance of ± 5%.

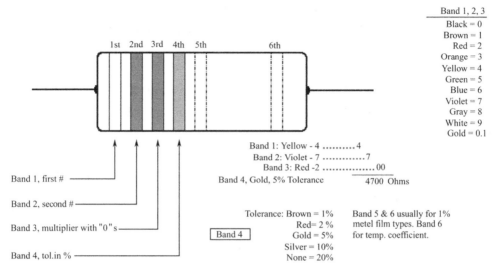

Fig 2.2   an example of resistor color code

For a 5-band resistor, the first band is the hundreds values, the second band gives the tens and the third band gives units, the forth band is a multiplying factor, the factor being 10's band value. The colors brown (1%), red (2%), green (0.5%), blue (0.25%), and violet (0.1%) are used as tolerance codes on 5-band resistors only. All 5-band resistors use a colored tolerance band.

Can you "create" your own resistors? Of cause and not difficult. Here is how to do it. Draw a line on a piece of paper with a soft pencil, HB or 2HB will do fine. Make the line thick and about 2 inches (5 cm) long. With your multimeter, measure the ohm's value of this line by putting a probe on each side of the line; make sure the probes are touching the carbon from the pencil. The value would probably be around the 800 kΩ to 1.5 MΩ depending line. The resistance will drop considerably, if you erase some of it (length-wise obviously!). You can also use carbon with silicon glue and when it dries measure the resistance, etc.

2.1.2 Capacitors

A capacitor is an electrical device that can temporarily store electrical energy. Basically, a capacitor consists of two conductors (metal plates) separated by a dielectric insulating material (Fig 2.3(a)), which increases the ability to store a charge. The dielectric can be paper, plastic film, mica, ceramic, air or a vacuum. The plates can be aluminum discs, aluminum foil or a thin film of metal applied to opposite sides of a solid dielectric. The conductor- dielectric-conductor sandwich can be rolled into a cylinder or left flat, the symbols of capacitor are shown in Fig 2.3(b).

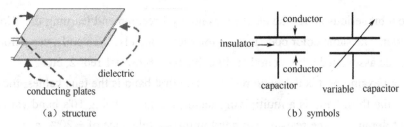

(a) structure　　　　　　　　　　　　(b) symbols

Fig 2.3 capacitor

A capacitor will block DC current, but appears to pass AC current by charging and discharging. It develops an AC resistance, known as capacitive reactance, which is affected by the capacitance and AC frequency. The formula for capacitive reactance is $X_C=1/(2\pi f_C C)$, with units of ohms.

Capacitors are available in various shapes and sizes (Fig 2.4). Usually, the value of capacitance and the working DC voltage are marked on them, but some types use a color code similar to resistors. Small-value capacitors of mica and ceramic dielectrics are indicated in pico farads (10^{-12} F), but only the significant digits are shown on the package, for example, '105' (Fig 2.4(d)) means 10×10^5 pF=1 μF. Tuning capacitors (such as used in radio) use air as a dielectric, with one set of plates, which can be rotated in and out of a set of stationary plates. Trimmer capacitors are used for fine adjustment with a screw, and have air, mica and ceramic as dielectrics.

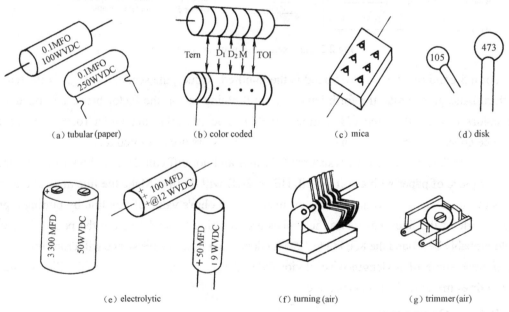

Fig 2.4 various capacitors

2.1.3 Inductors

An inductor is an electrical device, which can temporarily store electromagnetic energy in the field about it as long as current is flowing through it. The inductor is a coil of wire that may have an air core or an iron core to increase its inductance. A powered iron core in the shape of a cylinder may be adjusted in and out of the core.

An inductor tends to oppose a change in electrical current, it has no resistance to DC current but has an AC resistance to AC current, known as inductive reactance, this inductive reactance is affected by inductance and the AC frequency and is given by the formula $X_L=2\pi f_L L$, with units of ohms. Inductors are used for filtering AC current, increasing the output of the RF (radio frequency) amplifier.

Inductors are available in variety of shapes (Fig 2.5): air core, iron core (which may look like a transformer, but has only two leads), toroidal (doughnut shaped), small tubular with epoxy, RF choke with separate coils on a cylinder, and tunable RF coil with a screwdriver adjustment.

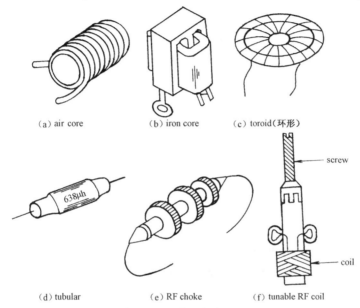

Fig 2.5　various inductors

Technical Words and Phrases

capacitance	[kəˈpæsɪtəns]	n. 电容量
capacitor	[kəˈpæsɪtə(r)]	n. (=capacitator) 电容器
charge	[tʃɑːdʒ]	n. 负荷，电荷，充电　v. 装满，充电
dielectric	[daɪɪˈlektrɪk]	n. 电介质，绝缘体　adj. 非传导性的
electrical	[ɪˈlektrɪkəl]	adj. 电的，有关电的
formula	[ˈfɔːmjulə]	n. 公式，规则，客套语
inductance	[ɪnˈdʌktəns]	n. 电感量，电感值
inductor	[ɪnˈdʌktə(r)]	n. 诱导物，感应器，电感器
insulating	[ɪnˈsjuleɪtɪŋ]	adj. 绝缘的
multimeter	[mʌltɪˈmiːtə(r)]	n. 万用表　vt. 多点测量
probe	[prɔb]	n. 探针，外太空探测器；探测飞船（=space probe）探索

reactance	[rɪˈæktəns]	n. 电抗
resistance	[rɪˈzɪstəns]	n. [电]电阻值，中文中有时简称电阻
resistor	[rɪˈzɪstə(r)]	n. [电]电阻器，中文中有时简称电阻
voltage	[ˈvəʊltɪdʒ]	n. [电工]电压，伏特数；电压（电位差）

be inversely proportional to	与……成反比
be affected by …	受……影响
be available	是可利用的，可用的
be directly proportional to	与……成正比
be expressed as a formula	用公式表示成……
be proportional to	与……成比例
be used to…	被用于……
doughnut shaped	环状的（doughnut 是指面包圈，其形状是圆环形）
in various shapes and sizes	各种形状和尺寸
to serve as…	用做……

2.2　Reading Materials

2.2.1　Nonlinear Resistors

Standard-type resistors usually maintain their value regardless of external conditions, such as voltage, temperature, and light. These types of resistors are referred to as linear resistors. There are other types of resistors referred to as nonlinear, whose resistance varies with temperature (thermistor), voltage (varistor) and light (photoresistor).

The thermistor is made from metal oxides, such as manganese, nickel, copper, or iron. Usually, a thermistor has a negative temperature coefficient, where an increase in temperature causes a decrease in its resistance. The typical resistance change is about –5%/ °C with a range of from 1 Ω to more than 50 MΩ. A thermistor might be used to control the stability of a transistor by being part of the biasing network. The thermistor is mounted close to the transistor. When the temperature increases, its resistance decreases. These results in less forward bias voltage from emitter to base; the current through the transistor decreases; and the circuit becomes more stable. When the temperature decreases, the thermistor resumes its initial value and the normal bias voltage is again present.

Varistors are similar in appearance to thermistors, but their resistance decreases with an increase in voltage. The current that flows in a varistor varies exponentially(U^n) with the applied voltage and may increase as much as 64 times for a given varistor. Most often, varistors are used as protection devices for other circuits, such as being placed in parallel across switch contacts to prevent sparking and in inductive circuits to prevent voltage surges.

A photoresistor (Fig 2.6) is made of a high resistance semiconductor. If light falling on the device is of high enough frequency, photons absorbed by the semiconductor give bound electrons enough energy to jump into the

Fig 2.6　photoresistor

conduction band. The resulting free electron (and its hole partner) conduct electricity, thereby lowering resistance.

2.2.2 Transformers

Transformers are related to the properties of inductors. If an inductor with an AC voltage across it is placed in parallel to a second inductor, the electromagnetic field of the first inductor will induce an AC voltage into the second inductor. When the two coils referred to as windings are placed on a core, they become a transformer. The input voltage is to the primary winding and the induced voltage is taken off the secondary winding. If the number of turns on the secondary is greater than the number of turns on the primary, the secondary voltage will be greater than voltage on the primary, referred to as a step-up transformer. If the number of turns on the secondary is less, the voltage is less, referred to as a step-down transformer.

There are many configurations to a transformer (Fig 2.7). There are air core (Fig 2.7(b)), iron core (Fig 2.7(a)) (increase the mutual inductance between windings) and phased windings (Fig 2.7(c)), since the secondary voltage is 180° out of phase with the primary voltage. An autotransformer has only three leads (Fig 2.7(d)) and one is c (common) connection. Some transformers have center-tapped secondary (Fig 2.7(e)) and multiple secondaries (Fig 2.7(f)) to develop various voltages. They are also available in various voltage ratings, sizes, and shapes.

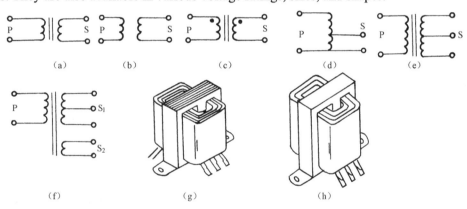

Fig 2.7 transformers

Transformers are used to step-up or step-down voltage for power supplies, for electrical isolation, for impedance matching, and couple signals between circuits.

Normally, DC is not useful on a transformer unless it be varying and it can saturate the windings. Therefore, a transformer is usually considered an AC device.

2.2.3 Various Low Voltage Apparatus

（以下是摘自某网站的低压电器的产品介绍，叙述简洁，却给出了产品的主要特征，通过阅读我们可以学会常用电器的描述方法。）

1. Motor Starters

NS2-25~80B AC motor starter is applicable to a circuit with AC voltage up to 690 V and current up to 80 A. It can be used to protect a three-phase cage asynchronous motor and a distribution line against overload, phase-failure and short circuit, to control the motor's infrequent starting and other infrequent load conversion. It can also serve as isolator.

2. QJ Series of Auto-reduced Voltage Starter

Auto-reduced voltage starter is applicable to infrequent reduced-voltage start up and stop of AC 50 Hz three-phase squirrel-cage induction motor with voltage of 380 V and power of 10, 14, 28, 40, 55 and 75 kW in general industry. When the tapped autotransformer starts, it reduces the power supply voltage, so that the starting current reduces. This product also provides with thermal relay and under-voltage tripping gear. When the motor is overloaded or the voltage of the line is at a certain value below the rated voltage, the motor will be switched off from the power supply to protect the motor.

3. CJX1 Series Contactor

CJX1 series AC contactor is mainly applicable to a line of AC 50 Hz or 60 Hz, rated insulation voltage 690 V~1 000 V, especially for application class AC-3 at the rated working voltage 380 V, rated working current 9 A~630 A. It is used to remotely close or/and switch off the line and to control start, stop and reverse-rotation of an AC motor.

4. HZ5 Series Combination Switch

HZ5 series packet type switches can mainly be applied in the electric circuits of AC 50 Hz, with voltage up to 380 V and below to be the controller of power supply switch and cage-squirrel inductive electromotor starting, conversion, and speed change switching, beside it can be applied to be connection of control circuits. The products are in conformity with GB 14048.3-93 standards.

5. DS30 Series Time Relay

It is used as an auxiliary element in various kinds of protection and automatic equipment to make the action of the controlled element to obtain required time delay, realizing selective combination of main protection and back-up protection.

6. NA1 Series Universal Circuit Breaker

NA1 series air circuit breaker (hereinafter referred to as breaker) is suitable for the circuit of AC 50 Hz with rated voltage 400 V, 690 V and rated current up to 6 300 A. It is mainly used to distribute electric energy and protect circuits and power-supply equipment from over-load, under-voltage, short-circuit and single-phase earthing. With intelligentized and selective protection functions, the breaker can improve the reliability of power supply, and avoid unnecessary power failure. The breaker is applicable for power stations, factories, mines (for 690 V) and modern high-buildings, especially for the distribution system of intelligentized building.

7. HG1 Series Fuse Type Isolator

HG1 series fuse type isolator (hereinafter called isolator) is applicable to power distribution circuit and motor circuit with rated AC voltage up to 380 V (50 Hz), conventional heating current up to 63 A and high short-circuit current as power supply isolator and circuit protection.

2.3 Knowledge about Translation
（翻译知识2——单词）

英语中的单词大多数都是多义词，应注意准确选定英语单词在句中的含义。根据文章涉及的专业内容来确定其含义是一种有效的方法。

1. 单词辨析

① charge

service charge	服务费
in charge of	负责……
furnance charge	炉料
induced charge	感应电荷

② develop

economic development	经济发展
intellectual development	智力开发
series development	级数展开
chemical development	化学显影（冲胶卷）

③ power

The sixth power of two is 64.	2的6次方是64。power在数学中为乘方，幂
horse power	马力
rated power	额定功率
1000-power microscope	千倍显微镜
power network	电力网
power switch	电源开关
power export	输出功率

在阅读电子技术专业文献时，应注意扩大自己的专业词汇。许多你平时很熟悉的单词，在电子专业文献中可能有其特定的含义。

2. 词义引申

在阅读理解过程中，有时直译的表达方式并不符合中文的表达习惯，这时应采用词义引申。在保持原文意思不变的情况下，用适当的引申词可以更准确地表达原文意思且符合中文表达习惯。

The light in the workshop is **poor**. 车间中光线不足。

Radar was not invented until **the last war**. 直到**上次战争**时，雷达还未发明。

The last war 指什么，在英语中是不言而喻的，但在汉语中就觉得意义不清，根据汉语习惯，应译成"雷达是在第二次世界大战中发明的"。

3．词量增删

由于历史背景、地理位置、自然环境、民情风俗等方面的巨大差异，英、汉两种语言的用词、结构和表达方式不可能完全相同，阅读时应该随时注意两种语言的差别，有时要增补一些词，有时要删减一些词，以符合中国人的阅读习惯。

① 词量增补。英语专业文章为了避免用词重复，常常省略一些词语，这是阅读时的难点之一。还有一些词，在英语中并无含义，但译成中文时应增补，方能正确理解。

If you are a bit serious about the electronics hobby I recommend the "Color Code". 如果你对电子技术颇有兴趣，建议学会"条形码"**电阻的识别方法**。

He offered to help us. 他提出**要**帮助我们。

Don't use more material than necessary. 要**多少**材料用多少，不要多用。

Let those who can serve as teachers. 让那些能当老师**的人**来当老师（能者为师）。

② 词量删简。英语中常用一些关系词、冠词、连词、代词以及同义词、同位语等，在译成汉语时可以免去不译。

It takes only three working processes to machine **the** part. 加工这个零件只要3道工序。（the 表示特指，故译成这个） It 这里不必译出。

The sixties of 18th century **saw the** start of **the** Industrial Revolution. 18世纪60年代开始了产业革命。

Multidigit displays consist of two or more seven-segment displays contained in a single package **or module**. 多位数码显示器由封装在一起的两个或更多的7段显示器组成（or module 为同义词，可不译）。

4．词类转换

有时中、英文在描述一些现象时用的方法是很不相同的，在专业文献中常常会遇到如果直译的话意思很不明确，这时可结合专业知识把原文的意思表达出来，把原来的词的词性进行转换，有时更符合中文的习惯。

How fast a machine **works** is one of its important characteristics. **运转速度**是机器的重要性能指标之一。

2.4 Exercises

1. Put the Phrases into English

① 电子元件　　　　　　　　⑥ 阻碍直流
② 欧姆定律　　　　　　　　⑦ 存储电能
③ 限制电流　　　　　　　　⑧ 感抗
④ 分压器　　　　　　　　　⑨ 绝缘材料
⑤ 晶体管偏置电路　　　　　⑩ 交流阻抗

2. Put the Phrases into Chinese

① known as capacitive reactance　　⑥ in the shape of a cylinder
② with units of ohms　　　　　　　⑦ block DC current, but pass AC current

③ prevent device from burning out
④ has an AC resistance to AC current
⑤ adjustment with a screw
⑧ to vary the inductance
⑨ be given by the formula
⑩ the RF amplifier

3. Sentence Translation

① Resistors are used to limit current flowing to a device, thereby preventing it from burning out, as voltage dividers to reduce voltage for other circuits, as transistor biasing circuits, and to serve as circuit loads.

② The resistor is generally a linear device and its characteristics form a straight line when plotted on a graph.

③ If you are a bit serious about the electronics hobby I recommend learning the "Color Code".

④ A capacitor will block DC current, but appears to pass AC current by charging and discharging.

⑤ Capacitors are available in various shapes and sizes.

⑥ A powered iron core in the shape of a cylinder may be adjusted in and out of the core to vary the inductance.

⑦ Capacitors are used for filtering, by passing signals, for timing circuits, and for radio-frequency (RF) tuning circuits.

⑧ An inductor is an electrical device, which can temporarily store electromagnetic energy in the field about it as long as current is flowing through it.

⑨ The inductive reactance is affected by inductance and the ac frequency.

⑩ The input voltage is to the primary winding and the induced voltage is taken off the secondary winding.

4. Write Main Clause of the Following Sentences

① However, to see how a circuit responds to a regular or repetitive input-the steady-state analysis - function that is by far the most useful is the sinusoid.

② The loudness of the information at the receiver is a result of the percentage of modulation.

③ In traversing any loop in any circuit, at every instant of time, the sum of the voltages having one polarity equals the sum of the voltages having the opposite polarity.

④ Finding the magnitude and phase angle of a sinusoidal steady-state response can be accomplished with either real or complex sinusoids.

⑤ Using the concepts of phasor and impedance, sinusoidal circuits can be analyzed in the frequency domain in a manner analogous to resistive circuits by using the phasor versions of KCL, KVL, nodal analysis，mesh analysis and loop analysis.

2.5 课文译文

2.5.1 电阻器

电阻器是一种电子元件，它能阻碍电流的流动。在电阻器中流过的电流与加在电阻两端的电压成正比，与电阻的阻值成反比。这就是欧姆定律，可以用公式表示成 $I=\dfrac{U_R}{R}$。电阻器一般是线性元件，它的（伏安）特性曲线形成一条直线。

电阻器常用做限流器，限制流过元件的电流以防止元件因流过的电流过大而烧坏。电阻器也可用做分压器，以减小其他电路的电压，如晶体管偏置电路。电阻器还可用做电路的负载。

一般来说，电阻有碳（膜）电阻、线绕电阻和金属膜电阻（如图 2.1 所示），电阻器尺寸的大小与电阻的（额定）功率有关，尺寸比较大的电阻器通常是高功率电阻器。可变电阻器是电阻值可调节的电阻器，如变阻器、电位器和微调电位器。精密电阻器是指其误差率在 1% 或更小的电阻器。

如果你对电子技术颇有兴趣，建议学会"彩色条形码"电阻的识别方法，这样会带来很多方便。而且这种彩色条形码标注方法在其他器件也适用，如线圈（电感）、电容等。每条彩条表示一个数字，如黑色=0，棕色=1，红色=2 等。

图 2.2 给出彩色条形码电阻的例子，这是一个有 4 色条码的电阻，其中第一条是十位数，第二条是个位数，第三条是以 10 为基的指数值，第四条表示电阻的精度。如果没有第四条，则电阻的精度为±20%，如果第四条为银色，表示电阻的精度为±10%，而金色表示精度为±5%。

对有 5 色条码的电阻来说，第一条是百位数，第二条是十位数，第三条是个位数，第四条是以 10 为基的指数值。褐色（精度 1%）、红色（2%）、绿色（0.5%）、蓝色（0.25%）和紫色（0.1%）只在 5 色条的电阻上用来表示精度。所有 5 色条的电阻都有表示精度的彩色条。

可以自己制作一个电阻（器）吗？当然可以而且不难。这里教你如何做一个电阻（器），用一支软铅笔（HB 铅笔或用 2HB 铅笔更好），在纸上画一条大约 2 英寸（5 cm）长的粗线。用万用表测量这段线的欧姆值，（方法是）把万用表的两个探笔分别与铅笔线的两端相接触，一定要让探笔与线段的碳接触。根据线的（粗细），电阻值大约在 800 kΩ～1.5 MΩ。如果你擦掉一些线，使线明显变短，电阻值就会变小。你也可以用含硅胶的碳来制作电阻器，当硅胶干了以后测量其电阻值。

2.5.2 电容器

电容器是可以暂时存储电能的电子元件。电容器一般由两块导体（金属极板）组成（见图 2.3(a)），中间用一层不导电的绝缘材料隔开，这层绝缘材料可以增加电容存储电荷的本领（即增大电容量）。绝缘材料可以是纸、塑料片、云母、陶瓷材料、空气或真空。极板可能是铝薄板，铝箔或在一片绝缘板的两面各贴上一层金属薄膜。可以直接把一个这种导体-绝缘体-导体（三明治式）制成平板电容器也可以把它卷起来成为圆柱形电容器，电容的符号如图 2.3(b)所示。

电容器隔直流，但能以充电和放电的方式通过交流。它构成的交流电阻抗，称为容抗。容抗与电容量和交流电的频率有关，容抗的公式为 $X_C=1/(2\pi f_C C)$，其单位为欧姆。

电容器有各种形状，大小不一（如图 2.4 所示）。通常电容值和加在电容两端的直流工作电压值是标在电容器上面的，但有些电容值和电阻值一样，是彩色条形码（来表示的）。用云母和陶瓷作为电介质的小电容器以皮法拉（10^{-12} F）作为电容单位，但在电容器上只印上有效值，如'105'（图 2.4(d)）表示其电容为 10×10^5 pF=1 μF（1 微法拉）。旋转电容器（如收音机中所用的）是用空气作为电介质的，由一组静止的平行电极板和一组可转动的电极板组成，当可转的电极板转进或转出时，电容量增大或减小。微调电容器用螺丝来进行精确调节，其绝缘材料有空气介质、云母介质和陶瓷介质。

2.5.3 电感器

当电流流过电感器时,电感器周围就有电磁场,电感器是以电磁场的形式暂时存储电磁能量的电子元件。电感器是一组线圈,有的电感器是空心的(空气芯),有的线圈中有可增加其电感量的铁芯,(可调电感)有一个强磁的圆柱状铁芯,通过调节铁芯可以增加电感量或减少电感量。

电感器总是反抗电流变化,对直流电而言,电感器是没有阻碍作用的,但对交流电来说,电感器有一个交流阻抗,称为感抗。这个感抗与电感量和交流电的频率有关,可以用公式表示为 $X_L=2\pi f_L L$,其单位为欧姆。电感器可以用来滤波,增加射频(无线电频率)放大器的输出。

电感器有各式各样的形状(如图 2.5 所示),空气芯的、铁芯的(铁芯的有时看起来像个变压器,但只有两个输出端)、环状的(圆环形的)、管状的,在一个圆柱体上有一些分开的线圈构成的射频扼流线圈和带有调节螺丝的可调射频线圈等。

2.6 阅读材料参考译文

2.6.1 非线性电阻器

标准类型的电阻器阻值通常不受外部条件如电压、温度和光的影响,这种电阻器称做线性电阻器。而非线性电阻器的阻值则随温度(热敏电阻器)、电压(压敏变阻器)或光(光敏电阻器)的变化而变化。

热敏电阻器是用金属如锰、镍、铜或铁的氧化物物制成的。通常,热敏电阻器有一个负的温度系数,当温度升高时它的阻值下降。典型的阻值变化约是−5%/℃,电阻的阻值范围为 1Ω~50MΩ。热敏电阻可以放在晶体管的偏置电路中来控制稳定晶体管电路的(静态工作点)。把一个热敏电阻安放在晶体管边上(并连接在电路中),当温度升高时,热敏电阻器的阻值下降,这就导致了从发射极到基极的正向偏置电压的减小,从而流过晶体管的电流减小,电路就变得比较稳定。当温度减小时,热敏电阻器恢复其初值,正向偏置电压也就恢复为正常值。

压敏变阻器在外形上与热敏电阻器相似,但它们的两端的电压增大时阻值减小,流过压敏变阻器的电流随其两端的电压按指数关系变化。对一个给定的压敏变阻器来说,电流可以增大到原来的 64 倍。大部分压敏变阻器都用做其他电路的保护器件,如并联在一个开关的两端以防止电火花或连接在一个感应电路中防止突然产生的电压波动。

光敏电阻(如图 2.6 所示)是用高阻的半导体材料制作的,如果频率足够高的光照在光敏电阻上,则半导体吸收光子(能量),使束缚电子能够跃迁到导通带而成为自由电子,自由电子(和相应产生的空穴)可以导电,则光敏电阻的阻值变小了。

2.6.2 变压器

变压器与电感器的特性有关。如果一个两端加有交流电的电感器以平行地靠近另一个电感器,第一个电感器产生的电磁场将在第二个电感器中感应产生一个交流电压。放在一个铁芯上两个线圈称为绕组,它们构成一个变压器。输入电压的称为第一(原)绕组,输出感应电压的称为第二(副)绕组。如果第二绕组上的线圈数大于第一绕组上的线圈数,第二绕组的电压将大于第一绕组的电压，这时称为升压变压器。如果第二绕组绕的圈数比较少，则输出的电压低，称为降压变压器。

变压器的结构有很多种（如图 2.7 所示），有空气芯的（图 2.7(b)）、铁芯的（图 2.7(a)）（增加绕组之间的电感量）和相位绕组的（图2.7(c)），即它的第二绕组与第一绕组的输出电压有 180°相位差。一个自耦变压器（图 2.7(d)）只有三个输出端，其中一个是 c（公共端）。有些变压器的第二绕组是中心抽头（图 2.7(e)）或多个副绕组（图 2.7(f)），以输出各种电压。变压器有各种额定电压级的，大小和形状也不一样。

变压器通常用于电源的升压、降压，电气隔离，阻抗匹配和电路之间的信号耦合。

一般来说，直流电（除非它是变化的）是不能输入变压器的，它会使线圈饱和，所以一个变压器通常认为是一个交流器件。

2.6.3 各种低压电器设备

1. 电动机启动控制器

NS2-25~80B 交流电动机启动器适用于交流电压 690 V 以下，交流电流 80 A 以下的电路，可用于三相鼠笼形异步电动机和配电线路的过载保护、失相保护和短路保护，用于控制电动机的不频繁启动和其他负载的不频繁切换，也可以用做电路隔离器。

2. QJ 系列自动减压启动器

自动减压启动器用于工业中常用的电压 380 V，功率 10 kW、14 kW、28 kW、40 kW、55 kW 和 75 kW，交流 50 Hz 三相鼠笼形感应电动机的频繁减压启动和停止（或译制动）。启动时自动分级接入自耦变压器，降低电压，从而减小启动电流。这个产品还带有热继电器和欠压释放机构，当电动机过载或线路电压低于电动机额定电压一定值时，会自动切除电动机的电源，以保护电动机。

3. CJX1 系列接触器

CJX1 系列交流接触器主要用于交流 50 Hz 或 60 Hz、额定绝缘电压 690～1000 V 的线路，尤其对额定（工作）电压 380 V、额定工作电流 9～630 A 的 AC-3 级电路，可以实现远距离接通、切断线路和控制一个交流电动机的启动、停止和反转。

4. HZ5 系列组合开关

HZ5 系列封装式开关主要用于交流 50 Hz、380 V 电压以下的电路，作为电源的控制开关和鼠笼形感应电动机启动、切换和调速开关，此外也可以用于控制电路的连接。本产品符合 GB 14048.3—93 标准。

5. DS30 系列时间继电器

时间继电器是各种保护电路和自动化设备的辅助控制电器，用于使控制电路按设定的时间延时动作，实现主电路保护和后备电路保护所选定的组合功能。

6. NAI 系列通用电路断路器

NAI 系列空气电路断路器（文中称为断路器）适用于交流 50 Hz，额定电压为 400 V/690 V 和额定电流 6 300 A 以下的电路，它主要用于分配电能和保护电路和电力设备，具有过载、欠压、短路和单相接地保护。这种具有智能化的选择性保护功能的断路器可以增强电源供应的可靠性，避免不必要的电力故障。这种断路器适用于发电站、工厂、矿业（690 V 的断路器）和高楼大厦（的配电线路），尤其是智能化大楼的配电系统。

7．HGI 系列熔断器型隔离器

HGI 系列熔断器型隔离器（文中称为隔离器）适用于额定交流电压 380 V（50 Hz）以下，通常过载电流在 63 A 以下，并具有很大的短路电流的电力配电线路和电动机电路，用做电源隔离器和电路保护。

Unit 3　Instruments

> *Pre-reading*
>
> Read the following passage, paying attention to the question.
> 1) What can be measured with a multimeter?
> 2) What is the function of a oscilloscope?

3.1　Text

3.1.1　Multimeters

A multimeter is a general-purpose meter capable of measuring DC and AC voltage, current, resistance, and in some cases, decibels. There are two types of meters: analog, using a standard meter movement with a needle (see Fig 3.1(a)), and digital, with an electronic numerical display (see Fig 3.1(b)). Both types of meters have a positive (+) jack and a common jack (−) for the test leads, a function switch to select DC voltage, AC voltage, DC current, AC current, or ohms and a range switch for accurate readings. The meters may also have other jacks to measure extended ranges of voltage (1 to 5 kV) and current (up to 10 A) there are some variations to the functions used for specific meters.

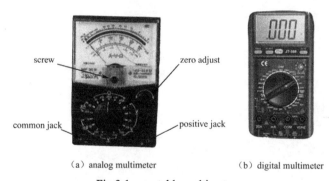

(a) analog multimeter　　　(b) digital multimeter

Fig 3.1　portable multimeters

Besides the function and range switches (sometimes they are in a single switch.), the analog meter may have a polarity switch to facilitate reversing the test leads. The needle usually has a screw for mechanical adjust to set it to zero and also a zero adjust control to compensate for weakening batteries when measuring resistance. An analog meter can read positive and negative voltage by simply reversing the test leads or moving the polarity switch. A digital meter usually has an automatic indicator for polarity on its display.

Meters must be properly connected to a circuit to ensure a correct reading. A voltmeter is always placed across (in parallel) the circuit or component to be measured. When measuring current, the circuit must be opened and the meter inserted in series with the circuit or component to be measured. When measuring the resistance of a component in a circuit, the voltage to the circuit must be removed and the meter placed in parallel with the component.

3.1.2 the Oscilloscope

The oscilloscope (Fig 3.2) is basically a graph-displaying device - it draws a graph of an electrical signal. When the signal inputs into the oscilloscope, an electron beam is created, focused, accelerated, and properly deflected to display the voltage waveforms on the face of a cathode-ray tube (CRT).

Fig 3.2   dual-trace oscilloscope

In most applications the graph shows how signals change over time: the vertical (*Y*) axis represents voltage and the horizontal (*X*) axis represents time. The amplitude of a voltage waveform on an oscilloscope screen can be determined by counting the number of centimeters (cm), vertically, from one peak to the other peak of the waveform (Fig 3.3) and the multiplying it by the setting of the volts/cm control. As an example, if the amplitude was 5 cm and the control was set on 1 V/cm, the peak-to-peak voltage would be 5 V.

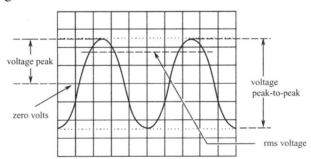

Fig 3.3   voltage peak and peak-to-peak voltage

Time can be measured using the horizontal scale of the oscilloscope. Time measurements include measuring the period, pulse width, and frequency. Frequency is the reciprocal of the period, so once you know the period, the frequency is one divided by the period.

The frequency of a waveform can be determined by counting the number of centimeters, horizontally, in one cycle of the waveform and the multiplying it by the setting time/cm control. For example, if the waveform is 4 cm long and the control is set at 1 ms/cm, the period would be 4 ms. The frequency can now be found from the formula:

$$f = \frac{1}{p} = \frac{1}{4\text{ms}} = 250 \text{ Hz}$$

If the control was gone on 100 μs/cm, the period would be 400 μs and the frequency would be 2.5 kHz.

A dual-trace oscilloscope is advantageous to show the input signal and output signal of one circuit in the same time, to determine any defects, and indicate phase relationships. The two traces may be placed over each other (superimposed) to indicate better the phase shift between two signals.

Technical Words and Phrases

accelerate	[əkˈseləreɪt]	v. 加速，促进
amplitude	[ˈæmplɪtjuːd]	n. 广阔，丰富，振幅
beam	[ˈbiːm]	n. 梁，（光线的）束，电波　v. 播送
compensate	[ˈkɔmpenseɪt]	v. 偿还，补偿，付报酬
decibel	[ˈdesɪbel]	n. 分贝
defect	[dɪˈfekt]	n. 缺点（故障，缺乏）
deflect	[dɪˈflekt]	v. （使）偏斜，（使）偏转
focus	[ˈfəukəs]	n. 焦点，焦距　vi. 聚焦　vt. 使集中在焦点上
horizontally	[hɔlɪˈzɔntlɪ]	adv. 水平地
jack	[ˈdʒæk]	n. 插孔，插座，起重器，千斤顶　vt.（用起重器）抬起
measure	[ˈmeʒə(r)]	n. 量度器，量度标准，测量　vt. 测量　vi. 量
meter	[ˈmiːtə(r)]	n. 米，公尺，仪表
needle	[niːˈd(ə)l]	n. 针（栅条，放射性材料容器）　vt. 刺穿
oscilloscope	[əˈsɪləskəup]	n. 示波器
period	[ˈpɪərɪəd]	n. 时期，学时，周期
superimposed	[sjuːpərɪmˈpəuzd]	adj. 成阶层的，有层理的，重叠的
switch	[swɪtʃ]	n. 开关，电闸，转换　vt. 转换，转变
vertically	[ˈvɜːtɪk(ə)lɪ]	adv. 垂直地

be connected to	连接到……
capable of V-ing	可以做……的，可以……的
cathode-ray tube (CRT)	阴极射线显像管
dual-trace oscilloscope	双踪示波器
in parallel with	与……并联
in series with	与……串联
Peak-to-peak voltage	电压峰–峰值
phase shift	相位漂移，移相
polarity switch	极性开关
rms voltage	电压有效值（rms 为 root-mean-square 的缩写）

3.2 Reading Materials

3.2.1 How Does an Oscilloscope Work

To better understand the oscilloscope controls, you need to know a little more about how oscilloscopes display a signal.

When you connect an oscilloscope probe to a circuit, the voltage signal travels through the probe to the vertical system of the oscilloscope. Fig 3.4 is a simple block diagram that shows how an analog oscilloscope displays a measured signal.

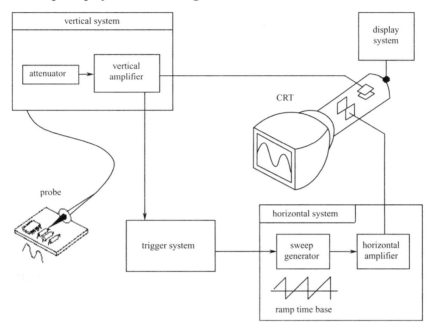

Fig 3.4　analog oscilloscope block diagram

Depending on how you set the vertical scale (volts/div control), an attenuator reduces the signal voltage or an amplifier increases the signal voltage.

Next, the signal travels directly to the vertical deflection plates of the cathode ray tube (CRT). Voltage applied to these deflection plates causes a glowing dot to move. (An electron beam hitting phosphor inside the CRT creates the glowing dot.) A positive voltage causes the dot to move up while a negative voltage causes the dot to move down.

The signal also travels to the trigger system to start or trigger a "horizontal sweep". Horizontal sweep is a term referring to the action of the horizontal system causing the glowing dot to move across the screen. Triggering the horizontal system causes the horizontal time base to move the glowing dot across the screen from left to right within a specific time interval. Many sweeps in rapid sequence cause the movement of the glowing dot to blend into a solid line. At higher speeds, the dot may sweep across the screen up to 500 000 times each second.

Together, the horizontal sweeping action and the vertical deflection action traces a graph of the signal on the screen. The trigger is necessary to stabilize a repeating signal. It ensures that the sweep begins at the same point of a repeating signal, resulting in a clear picture as shown in Fig 3.5.

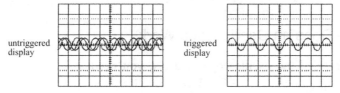

Fig 3.5   triggering stabilizes a repeating waveform

In conclusion, to use an analog oscilloscope, you need to adjust three basic settings to accommodate an incoming signal:

1. the Attenuation or Amplification of the Signal

Use the volts/div control to adjust the amplitude of the signal before it is applied to the vertical deflection plates.

2. the Time Base

Use the sec/div control to set the amount of time per division represented horizontally across the screen.

3. the Trigger Level of the Oscilloscope

Use the trigger level to stabilize a repeating signal, as well as triggering on a single event.

3.2.2   Analog and Digital Oscilloscope

Electronic equipment usually can be divided into two types: analog and digital. Oscilloscopes also come in analog and digital types. An analog oscilloscope works by directly applying a voltage being measured to an electron beam moving across the oscilloscope screen. The voltage deflects the beam up and down proportionally, tracing the waveform on the screen. This gives an immediate picture of the waveform.

In contrast, a digital oscilloscope samples the waveform and uses an analog-to-digital converter (or ADC) to convert the voltage being measured into digital information. It then uses this digital information to reconstruct the waveform on the screen.

For many applications either an analog or digital oscilloscope will do. However, each type does possess some unique characteristics making it more or less suitable for specific tasks.

People often prefer analog oscilloscopes when it is important to display rapidly varying signals in "real time" (or as they occur).

Digital oscilloscopes allow you to capture and view events that may happen only once. They can process the digital waveform data or send the data to a computer for processing. Also, they can store the digital waveform data for later viewing and printing.

3.2.3   Wattmeter

The traditional analog wattmeter (Fig 3.6) is an electrodynamic instrument. The device consists of a pair of fixed coils, known as current

Fig 3.6   wattmeter

coils, and a movable coil known as the potential coil.

The current coils connected in series with the circuit, while the potential coil is connected in parallel when measure with a wattmeter. Also, on analog wattmeters, the potential coil carries a needle that moves over a scale to indicate the measurement. A current flowing through the current coil generates an electromagnetic field around the coil. The strength of this field is proportional to the line current and in phase with it. The potential coil has, as a general rule, a high-value resistor connected in series with it to reduce the current that flows through it.

The result of this arrangement is that on a dc circuit, the deflection of the needle is proportional to both the current and the voltage, thus conforming to the equation $P=VI$. On an ac circuit the deflection is proportional to the average instantaneous product of voltage and current, thus measuring true power, and possibly (depending on load characteristics) showing a different reading to that obtained by simply multiplying the readings showing on a stand-alone voltmeter and a stand-alone ammeter in the same circuit.

A modern digital electronic wattmeter/energy meter samples the voltage and current thousands of times a second. The average of the instantaneous voltage multiplied by the current is the true power. The true power divided by the apparent volt-amperes (VA) is the power factor. A computer circuit uses the sampled values to calculate RMS voltage, RMS current, VA, power (watts), power factor, and kilowatt-hours. The simple models display that information on LCD. More sophisticated models retain the information over an extended period of time, and can transmit it to field equipment or a central location.

3.2.4 Signal Generator

A signal generator converts DC to AC or varying DC in the form of sine waves, square waves, triangle waves, or other types of voltage waveforms (Fig 3.7). The signal generator is used to inject a signal into a circuit or piece of equipment for troubleshooting or for calibration. Some generators may be used for audio, RF, or higher frequencies, whereas others have overlapping frequency ranges. All generators will have a function switch, a frequency range switch, and a fine adjustment control for selecting a specific frequency, an amplitude control for varying the peak-to-peak output voltage, and output terminals.

Fig 3.7 function signal generator

To select a sine wave of, say, 5 kHz, the user would set the function switch to the sine wave, set the range switch to 1 k, and then adjust the frequency fine adjust control to 5. The amplitude control would then be adjusted to establish the desired peak-to-peak voltage output.

3.3 Knowledge about Translation
（翻译知识 3——非谓语动词 V-ing）

前面提到，在英语的一句话中如果要叙述几个动作时，先选其中主要的动作作为谓语，其余动作改用非谓语动词，否则不符合英语的语法。非谓语动词通常有三种：V-ing，V-ed 和 to V。

V-ing 形式的动词在有的语法书中分为动名词与现在分词，但现在的趋势是不加区分。本书中不加区分，只要能准确理解句子中的意思没有必要去辨别它是动名词还是现在分词。V-ing 在句中可以作为表语，主语，宾语，定语，同时保留了动词性，因此可带有宾语和状语。

① V-ing 本身含有主动、进行的意思，表示动作是由所修饰的名词主动发出的。

The Zener diode maintains the voltage across its terminals by **varying** the current that flows through it. 稳压管通过**改变**流过它的电流来维持它两端电压的**稳定**（varying 表示稳压管主动变化，"稳定"为词语增补）。

The second step in **minimizing** costs of production is **choosing** the cheapest of the technical efficient alternatives. 降低生产成本的第二个措施是选用最便宜的技术上有效的替代品。

A multimeter is a general-purpose meter capable of **measuring** DC... 一个万用表是能**测量**直流……的通用仪表。

② V-ing 作定语。

单个 V-ing 作定语一般放在名词前面（也可以放在后面），V-ing 短语作定语一般放在名词之后。

All **moving** bodies have energy. 所有运动的物体都有能量。

A direct current is a current **flowing** always in the same direction. 直流电流是指流动方向不变的电流。

A zero adjust control to compensate for **weakening** batteries when measuring resistance. 当测量电阻时，零点调节控制（钮）可用于在电池电压不足时进行补偿调节（即保证电阻为 0 时指针指向零值）。

③ 作状语。

V-ing 短语作状语时，往往具有时间、条件、原因、结果、方式、补充说明等含义，它可放在句首、句中或句尾，通常它的逻辑主体就是句中的主语。

Lifting something, you do work. 当你举起某物时，你就在做功（条件状语）。

Being negative, electrons move always from negative to positive. 电子是负的，所以总是从负（极）向正（极）运动（V-ing 短语作原因状语）。

分词短语作状语时，前面可用 when, while, if, unless, though 等连词来加强时间、条件等含义。

When **measuring** current, the circuit must be opened and the meter inserted in series with the circuit or component to be measured. 当测量电流时，必须断开电路，以将万用表与待测电路或元器件相串联。

④ 作主语或宾语。

Heating the water changes it into vapor. 把水加热可以使水变为蒸汽（V-ing 短语作主语）。

It may also have a polarity switch to facilitate **reversing** the test leads. 还有一个极性开关可以用来很方便地交换测试笔的极性。（V-ing 短语作 facilitate 的宾语）

⑤ 作主语或宾语的补足语。

We put a hand above an electric fire and feel the hot air **rising**. 我们把手放在电炉的上方，就会感觉到热空气在上升。

⑥ 与 with (without) 连用：

在科技文章中，with (without) + 名词 + 分词 结构常用做补充说明。在这种结构中，with 没有词汇的意思，表示一种伴随情况，可根据具体情况进行理解。

The density of air varies directly as pressure, with temperature **being** constant. 在温度不变时，空气密度与压力成正比。

⑦ 有些动词后面只能接 V-ing 形式的动词。

这类动词主要有 bar（禁止），delay（推迟），avoid（避免），forbid(禁止), miss（错过，没赶上），save（免得）。

I have to delay answering the letter. 我只能推迟写回信。

有些动词后面习惯接 V-ing 形式的动词，用主动形式表示被动意义，这类动词主要有 bear（忍受），deserve（该，值得），require（需要），stand（忍受），want（要）等。

The machine wants repairing. 这台机器需要修了。

3.4 Exercises

1. Put the Phrases into English

① 通用仪表
② 模拟仪表
③ 交换测试笔
④ 机械调节
⑤ 测量电阻
⑥ 正向电压
⑦ 测量电流
⑧ 电压幅度
⑨ 双踪示波器
⑩ 信号发生器

2. Put the Phrases into Chinese

① analog multimeter
② extended range
③ specific meters
④ includes the function and range switches
⑤ present an electronic picture
⑥ display the voltage waveform
⑦ appear on the screen
⑧ phase relationships
⑨ as an example
⑩ in series with the circuit

3. Sentence Translation

① A multimeter is a general-purpose meter capable of measuring DC and AC voltage, current, resistance, and in some cases, decibels.

② An analog meter can read positive and negative voltage by simply reversing the test leads or moving the polarity switch.

③ When measuring current, the circuit must be opened and the meter inserted in series with the circuit or component to be measured.

④ The frequency of a waveform can be determined by counting the number of centimeters, horizontally, in one cycle of the waveform and the multiplying it by the setting time/cm control.

⑤ Depending on how you set the vertical scale (volts/div control), an attenuator reduces the signal voltage or an amplifier increases the signal voltage.

⑥ Lissajous patterns can be used to show the phase relationship of two signals of the same frequency and to determine an unknown frequency from a known frequency.

⑦ There are two types of meters: analog, using a standard meter movement with a needle, and digital, with an electronic numerical display.

⑧ If both signals are the same frequency, a circle will appear on the face of the oscilloscope.

⑨ In most applications the graph shows how signals change over time: the vertical (*Y*) axis represents voltage and the horizontal (*X*) axis represents time.

⑩ All signal generators will have a frequency range switch, a fine adjustment control for selecting a specific frequency, an amplitude control for varying the peak-to peak output voltage, and output terminals.

4. Translation

This book belongs in the technical library of anyone involved in the field of electronics. It provides the basic theory, components, devices, circuits, and systems of electronics. Not only does this book provide quick and accessible information, but it contains many practical suggestions together with basic testing and troubleshooting procedures.

3.5 课文参考译文

3.5.1 万用表

万用表是一种通用仪表，能用来测量直流和交流电压、电流、电阻，有的还能测量分贝（放大倍数）。有两种万用表：一种是用指针在标准刻度上的移动来指示测量值的模拟万用表（如图 3.1(a)所示），另一种是用电子数字显示器显示测量值的数字万用表（如图 3.1(b)所示）。这两种万用表都有一个正极（+）插孔和一个公共端（−）插孔用来插入测试笔，一个功能选择开关用来选择（测量对象）：直流电压、交流电压、直流电流、交流电流或欧姆（电阻），一个范围选择开关用来（选择范围以做）精确测量。万用表也可能还有其他插孔用来测量高电压（1~5 kV）和大电流（高达 10 A），对一些特殊的万用表来说还有一些其他功能的变化。

除了功能选择开关和范围选择开关（有时它们合成一个开关），模拟万用表可能还有一个极性开关，可以很方便交换测试笔的极性。通常有一个螺丝可调节指针，（在无电流时）使指针指在零处。当测量电阻时另有一个零点调节控制（钮）用来在电池电压不足时进行补偿调节（即保证电阻为 0 时指针指向零值）。一个模拟万用表可以测量正电压和负电压，只要简单地对调一下两个测试笔或拨一下极性开关。一个数字万用表通常会自动在显示器上指示出极性。

为了保证读数正确，万用表必须与电路正确连接。一个电压表（万用表测量电压时）应与被测电路或元件并联。当测量电流时，电路必须断开，插入万用表表笔使万用表与被测电路或元件相串联。当测量电路中局部电路（或元件）的（等效）电阻时，必须除去电路中的电源，万用表与这局部电路（或元件）并联。

3.5.2 示波器

示波器（如图 3.2 所示）是一个图像显示设备，它显示一个电子信号的图像。当信号输入到示波器中时，一个电子束被产生、聚焦、加速并适当偏离，在阴极射线管（CRT）的显示屏上显示电压的波形。

通常，示波器图像显示信号如何随时间变化：其垂直轴 Y 表示电压，水平轴 X 表示时间。在示波器屏幕上的电压波形的幅度可以通过数出电压波峰与波谷之间纵向距离的厘米值来确定（如图 3.3 所示），将这厘米值乘上 V/cm 控制钮的设定值就得到电压的幅度值。举例说，如果电压的峰-峰间幅度为 5 cm，控制钮设在 1 V/cm 处，则峰-峰电压值为 5 V。

用示波器的水平标尺可以测量时间值，时间测量包括测量信号的周期、脉冲宽度和频率。频率是周期的倒数，所以一旦知道了周期，频率就是用 1 除以周期。

一个波形的频率可以通过在水平方向数出波形一个周期的厘米值来确定，将这厘米值乘上时间/厘米控制钮的设定值就得到它的一个周期所需的时间。例如，如果一个波形长 4 cm，控制钮设在 1 ms/cm，则周期是 4 ms，则频率可以用下面的公式求出：

$$f = \frac{1}{p} = \frac{1}{4 \text{ ms}} = 250 \text{ Hz}$$

如果控制钮设在 100 μs/cm，则周期是 400 μs，频率为 2.5 kHz。

一个双踪示波器具有同时显示输入信号和输出信号的优点，可以显示输出信号是否有失真和表示输入/输出信号的相位关系。两路信号的波形可以重叠在一起，较好地显示出输入信号与输出信号相位的差别。

3.6 阅读材料参考译文

3.6.1 示波器是如何工作的

为了更好地掌握示波器的操作方法，你需要知道示波器是如何显示信号（图像）的。

当把示波器的探头与一个电路相连接时，电压信号通过探头送到示波器的垂直系统，图 3.4 是一个简单的框图，说明一个模拟示波器是如何显示一个测量信号的。

一旦设定了垂直单位（电压/单位格），就确定了是用衰减器还是用放大器（把输入信号转换成适当幅度的电压信号）。

接着，信号直接传到阴极射线管的垂直偏离板上，施加到这些偏离板的电压导致发光点的移动（一个电子束撞击阴极射线管内部磷（荧光物质）上产生光点），一个正电压使光点向上移动，而一个负电压使光点向下移动。

同时信号也输入到触发系统去启动或触发一个"水平扫描"，水平扫描是一个专业词汇，是指水平系统的作用导致光点沿着屏幕（水平）移动，水平系统的触发导致水平扫描发生器驱动光点在一个指定的时间间隔内从屏幕的左边移到右边。快速有序多次扫描导致这些光点的移动混合成一条实线。对高速扫描的示波器，每秒中扫过屏幕的光点可能高达 50 万次。

水平扫描和垂直偏离的作用合在一起就在屏幕上形成了信号的图像轨迹。为了稳定一个重复（周期性）的信号，触发是必要的，它保证在重复（周期性）信号的同（相位）点开始扫描，从而产生一个清晰的图像，如图 3.5 所示。

总之，在使用一个模拟示波器时，需要根据输入信号调节（下列）3 个基本的设置（钮）

使示波器显示适当的图像。

1. 信号的衰减或放大

用电压/单位格控制旋钮调整输入信号（到垂直偏离板之前的）幅度。

2. 时间基准

用秒/单位格设置旋钮设置屏幕上水平线每单位格所表示的时间。

3. 示波器触发控制旋钮

用触发对准（钮）来稳定一个重复（周期）信号或触发一个单（无周期）信号。

3.6.2 模拟示波器和数字示波器

电子设备通常都可以分成两类：模拟的和数字的。示波器也有模拟的和数字的。模拟示波器工作时是直接用待测电压去控制电子束在示波器屏幕上的运动，电压使电子束向上或向下偏离，偏离幅度与电压值成正比，在屏幕上显示波形。这种方法是给出波形的瞬时图像。

作为对比，数字示波器却是对电压波形采样，并用一个模/数转换器把待测电压转换成数字信号，然后用这个数字信号在屏幕上重新"构画"出波形。

在很多应用中，模拟示波器和数字示波器都可以测量显示，但是，每种（示波器）处理过程中有一些独特的特点，根据具体的测量对象，有的示波器可能更合适些。

当需要显示快速变化的"实时信号"时人们喜欢用模拟示波器。

数字示波器可以捕获图像并且一次测量后可以多次重复显示。数字示波器可以处理数字化的波形数据或把这些波形数据送到计算机中去处理。数字示波器还可以存储数字化的波形数据，（供）以后（再次）显示和打印。

3.6.3 瓦特表

传统的模拟瓦特表（如图 3.6 所示）是一种电动仪表。它由一个固定线圈（称为电流线圈）和一个可转动的线圈（称为电势或电压线圈）组成。

用瓦特表测量时，电流线圈与（待测）电路串联，而电压线圈与（待测）电路并联。模拟瓦特表的电压线圈还带一个指针，此指针可转动，用来在面板上指示测量值。流过电流线圈的电流在线圈周围产生一个电磁场，这个电磁场的强度与导线中的电流大小和相位成正比。通常，电压线圈与一个高阻值的电阻串联，以减小流过电压线圈中的电流。

这样设置的结果是在测量直流电路时，指针的偏转正比于电流和电压，符合 $P=VI$，在测量交流电路时指针的偏转则正比于电压和电流瞬时乘积的平均值，即测出有功功率。有时（取决于负载的性质）这个读数是不同于用电压表和电流表对同一电路进行测量的测量值之乘积（视在功率）的。

现在的数字电子瓦特表在一秒内对电压和电流进行数千次的采样，瞬时电压和电流积的平均值就是有功功率。有功功率除以视在功率（VA）就是功率因数。计算机电路根据所采样的值求出电压的有效值、电流的有效值、有功功率、功率因数和千瓦-小时（度）。简单的瓦特表直接在 LCD 上显示这些值。功能更强的瓦特表能保存一定时间段的信息并能把这些信息传给其他的设备或中心（控制设备）。

3.6.4 信号发生器

信号发生器是把直流电转换成交流电或变化的直流电的仪器（如图3.7所示），即把直流电转换成正弦波、方波、三角波或其他波形的电压信号。信号发生器是用来给电路或设备输入一个信号，以便对电路或设备进行维修或校正用的。有些信号发生器可以用来专门产生音频、射频或高频信号，而另一些则可以产生多种频率范围的信号。所有的信号发生器有函数（波形选择）旋钮，有频率范围选择旋钮、频率细调旋钮，用来选择一个特定的频率，有一个幅度控制旋钮用来改变输出电压的峰-峰值（或幅值），还有一些输出端口。

如果要选一个5 kHz的正弦波，可把波形选择旋钮调在正弦波上，把频率范围选择旋钮放在1 k上，然后调节频率细调旋钮到5，再通过调节输出幅度控制旋钮便可得到想要的峰-峰值电压的输出信号。

Unit 4 Electronic Components

Pre-reading

Read the following passage, paying attention to the question.
1) What is a semiconductor diode?
2) What difference is between the forward biased and reverse biased?
3) Can a transistor be used to amplify a signal?
4) What is a CMOS circuit?

4.1 Text

4.1.1 Semiconductor Diode

A semiconductor diode (refers to diode in short) is the simplest possible semiconductor device. A diode consists of a PN junction made of semiconductor material. The P-type material is called the anode, while the N-type material is called the cathode (Fig 4.1).

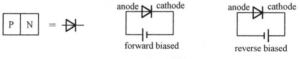

Fig 4.1 diode

A diode is forward biased when the anode is more positive than the cathode (greater than the turn-on voltage, which is approximately 0.3 V for germanium and 0.7 V for silicon). In this condition the internal resistance of the diode is low and a large current will flow through the diode (depending on the external circuit resistance).

The diode is reverse biased when the anode is less positive than the cathode. In this case, the internal resistance is extremely high, so perfect diodes can block current in one direction while letting current flow in another direction.

Diodes can be used in a number of ways. For example, a device that uses batteries often contains a diode that protects the device if you insert the batteries backward. The diode simply blocks any current from leaving the battery if it is reversed – this protects the sensitive electronics in the device.

A diode's behavior is not perfect, as shown in Fig 4.2. When reverse-biased, an ideal diode would block all current. A real diode lets perhaps 10 microamps through – not a lot, but still not perfect. And if you apply enough reverse voltage (V), the junction breaks down and lets current through. Usually, the breakdown voltage is a lot more voltage than the circuit will ever see, so it is irrelevant.

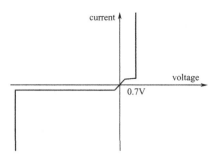

Fig 4.2　a diode's behavior

When forward-biased, there is a small amount of voltage necessary to get the diode going. In silicon, this voltage is about 0.7 volts. Though a large forward current can flow through the diode, too much current through the diode in either direction will destroy it.

Some pictures of diodes are given in Fig 4.3.

Fig 4.3　various diodes

4.1.2　NPN Bipolar Transistor

There are two types of standard bipolar transistors, NPN and PNP, with different circuit symbols (Fig 4.4). The letters refer to the layers of semiconductor material used to make the transistor. Most transistors used today are NPN because this is the easiest type to make from silicon.

The NPN bipolar transistor consists of an N-type emitter (E), P-type base (B), and N-type collector (C).

An amplifier can be built with a transistor. Fig 4.5 shows the two current paths through a transistor. You can build this circuit with two standard 5mm red LEDs and any general purpose low power NPN transistor.

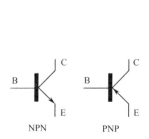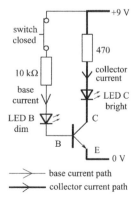

Fig 4.4　transistor circuit symbols　　　Fig 4.5　the small current controls the larger current

When the switch is closed, a small current flows into the base (B) of the transistor. It is just enough to make LED B glow dimly. The transistor amplifies this small current to allow a larger current to flow through from its collector (C) to its emitter (E). This collector current is large enough to make LED C light brightly.

The amount of collector current (I_C) is directly proportional to the amount of base current (I_B) and the collector current (I_C) will be less than the emitter current (I_E), since a small base current (I_B) must flow to turn on the transistor. The relationship of the currents is $I_E = I_C + I_B$. The ratio of I_C to I_B is called the current gain of the transistor and indicates its ability to amplify. This current gain is called beta (β) and is expressed as $\beta = \Delta I_C / \Delta I_B$, when the voltage from C to E (U_{CE}) is held constant.

To turn on an NPN bipolar transistor, the base must be more positive than the emitter (about +0.6 V for silicon). When the transistor is turned on hard (in saturation), this voltage is about +0.7 V and the resistance from C to E is low and may even appear almost as a short.

When the switch (Fig 4.5) is open no base current flows, so the transistor switches off the collector current and both LEDs are off. The resistance from C to E now is high and may appear as an open. Actually a transistor's behavior is not perfect too, a small leakage current (I_{CBO}) from C to B is always present and may cause stability problems for a transistor circuit.

The Fig 4.6 displays various bipolar transistors.

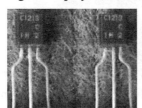  

Fig 4.6 various bipolar transistors

4.1.3 MOS Transistors

Presently, the most popular technology for realizing microcircuits makes use of MOS transistors. The acronym MOS stands for metal-oxide semiconductor, which historically denoted the gate, insulator, and channel region materials respectively. The MOS transistors can be divided into two classes in terms of carrier: N-channel and P-channel, electrons are used to conduct current in N transistors, whereas holes are used in P transistors While N devices conduct with a positive gate voltage, P devices conduct with a negative gate voltage.

The function of the MOS transistor is controlling a large current with a small voltage.

According to the voltage condition, there are two kinds of MOS transistor: enhancement and exhausted. The symbols used for enhancement MOS transistors and exhausted MOS transistors are shown in Fig 4.7. MOS transistors are actually four-terminal devices; the fourth terminal is a substrate connection. For digital circuits, the substrate connection of N-channel transistors is almost always the most negative IC voltage (i.e. ground or U_{SS}). Similarly, the substrate connection for P-channel transistors will be assumed to be the most positive IC voltage, which is labeled U_{DD}. This will always be assumed to be the case unless stated otherwise, and therefore substrate connections will not be shown.

Unlike most bipolar-junction transistor (BIJ) technologies, which make dominant use of only one type of transistor, MOS circuits normally use two complementary types of transistors. Microcircuits containing both N and P transistors are called CMOS circuits, for complementary MOS.

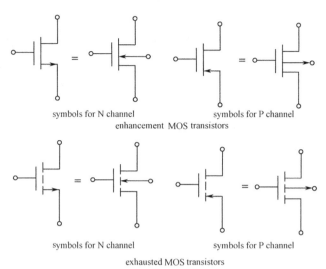

Fig 4.7   MOS transistors

4.1.4   Ideal Operational Amplifier

Of fundamental importance in the study of electric circuits is the ideal voltage amplifier or ideal operation amplifier (Op amplifier). Such a device, in general, has two inputs, v_1 and v_2, and one output, v_o. The relationship between the output and the inputs is given by $v_o=A(v_1-v_2)$, where A is called the gain of the amplifier. Note that since the ideal amplifier input resistance $R=\infty$, when such an amplifier is connected to any circuit, no current will go into the input terminals. Also, since the output v_o is the voltage across an ideal source, we have that $v_o=A(v_1-v_2)$, regardless of what is connected to the output for the sake of simplicity, the ideal amplifier having gain A is often represented as shown in Fig 4.8. We refer to the input terminal labeled "-" as the inverting input and the input terminal labeled "+" as the noninverting input.

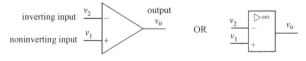

Fig 4.8   operational amplifier

Technical Words and Phases

acronym	['ækrənɪm]	n. 首字母简略词
anode	['ænəud]	n. [电]阳极，正极
biased	['baɪəst]	adj.偏压的，偏置的
bipolar	[baɪ'pəulə(r)]	adj.有两极的，双极性的
cathode	['kæθəud]	n. 负极，阴极

collector	[kə'lektə(r)]	n. （计算机）集电极，收藏家，征收者
complementary	[kɔmplə'mentərɪ]	adj. 补充的，补足的
diode	['daɪəud]	n. 二极管
dominant	['dɔmɪnənt]	adj. 占优势的，支配的，主要的
emitter	[ɪ'mɪtə(r)]	n. 发射体（发射极，辐射源）
enhancement	[ɪn'hænsmənt]	n. 增强（提高，放大），增进，增加
gain	[geɪn]	n. 增益，放大系数 vt. 得到，增进，赚到 vi. 获利；增加增益
germanium	[dʒə:'meɪnɪəm]	n. 锗
inverter	[ɪnvɜ:tə(r)]	n. 反相器
leakage	[lɪ:'kɪdʒ]	n. 漏，泄漏，渗漏
polysilicon	[pəlɪ'sɪlɪkən]	n. 多晶硅
silicon	['sɪlɪkən]	n. 硅
stability	[stə'bɪlɪtɪ]	n. 稳定性
substrate	['sʌbstreɪt]	n. (=substratum) 基底，底层，下层
symbol	['sɪmb(ə)l]	n. 符号，记号，象征
terminal	['tɜ:mɪn(ə)l]	n. 终端，接线端，终点站 adj. 末期

be assumed to	被假设成……
be labeled	被分类为……，被标志为……，被标注为……
break down	毁掉，制服，压倒，停顿（此处指二极管烧坏）
current operated device	电流控制器件
forward/reverse biased	正偏置/反偏置
internal resistance	内阻
microcircuit	微电路
Operational Amplifier	集成运算放大器，简写 Op-Amplifier 或 Op-Amp
inverting input	反相输入，指输出信号与此端输入的信号相位相反
noninverting input	同相输入，指输出信号与此端输入的信号相位相同

Notes to the Text

1. PN junction　　PN 结
2. N-channel transistor　　N 沟道晶体管
3. bipolar-junction transistor (BJT)　　双极性结式晶体管，中文常称晶体三极管

4.2 Reading Materials

4.2.1 Audio Amplifiers

For those of you who like to experiment with audio circuits and would like a simple amplifier that frees you from having to figure out the biasing resistors, we have two for you (and they run off

9 Volts too!). One uses an Op-Amp (Fig 4.9 (a)) and the other uses a transistor (Fig 4.9(b)).

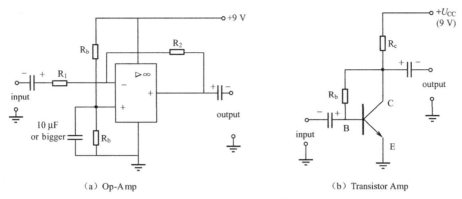

(a) Op-Amp (b) Transistor Amp

Fig 4.9   two audio amplifiers

Both circuits need capacitors on the input and output to block DC while passing AC. The capacitor values depend on which circuit you use and what signal frequency you are amplifying. Start with a 1 µF and go from there. Low frequencies may require a bigger value.

While Op-Amps normally run off of a dual voltage supply (+U and -U), it is possible to run them from a single voltage by using two equal value resistors (R_b) to create a separate DC grounding point midway between U_{CC} and actual ground, just for biasing the Op-Amp. The DC ground is connected to actual ground through a by-pass capacitor. The value of R_b is not critical; 10 k should work just fine. To minimize DC offset in the output, R_b should have a 1% tolerance.

The gain of the amplifier is set by R_1 and R_2 ($A_V = R_2/R_1$). R_2 should be at least 2k or bigger so as not to load the Op-Amp too much. If you use a bipolar device such as the venerable 741, the output can't go lower than 2 volts above ground or higher than 2 volts below U_{CC}. So with a 9-volt battery, the maximum output swing will be 5 volts: from 2 V to 7 V. If you want to go "rail-to-rail" from U_{CC} to Ground, then use a CMOS device like the CA3130; U_{CC} can then be as high as the Op-Amp allows. The CA3130 requires a 100 pF compensation capacitor.

If you really want "quick and dirty", this one transistor circuit is an "oldie but goodie". (See Fig 4.9 (b)). Note that by connecting the base-bias resistor R_b to the collector you get two benefits: 1) the biasing cannot cause saturation or cut-off and 2) you introduce some negative feedback into the signal path, which reduces distortion. It's not as good as the Op-Amp circuit but it does work. As for gain, you'll just have to measure it and see. Play around with different values of R_c, and make $R_b = 100R_c$. The voltage peak of input signal should not exceed 15 mV.

4.2.2   the Transistor as a Switch

Because a transistor's collector current is proportionally limited by its base current, it can be used as a sort of current-controlled switch. A relatively small flow of electrons sent through the base of the transistor has the ability to exert control over a much larger flow of electrons through the collector.

For the sake of illustration, let's insert a transistor in place of the switch to show how it can control the flow of electrons through the lamp (Fig 4.10). Remember that the controlled current through a transistor must go between collector and emitter. Since it's the current through the lamp that we want to control, we must position the collector and emitter of our transistor where the two contacts of the switch are now. We must also make sure that the lamp's current will move in the direction of the emitter arrow symbol to ensure that the transistor's junction bias will be correct.

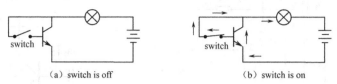

(a) switch is off   (b) switch is on

Fig 4.10   the transistor as a switch

In this example I happened to choose an NPN transistor.

We are faced with the need to add something more so that we can have base current. Without a connection to the base wire of the transistor, base current will be zero, and the transistor cannot turn on, resulting in a lamp that is always off.

If the switch is open, the base wire of the transistor will be left "floating" (not connected to anything) and there will be no current through it (Fig 4.10(a)). In this state, the transistor is said to be cutoff. If the switch is closed (Fig 4.10(b)), however, electrons will be able to flow from the emitter through to the base of the transistor, through the switch and up to the left side of the lamp, back to the positive side of the battery. This base current will enable a much larger flow of electrons from the emitter through to the collector, thus lighting up the lamp. In this state of maximum circuit current, the transistor is said to be saturated.

Of course, it may seem pointless to use a transistor in this capacity to control the lamp. After all, we're still using a switch in the circuit, aren't we? If we're still using a switch to control the lamp – if only indirectly – then what's the point of having a transistor to control the current? Why not just use the switch directly to control the lamp current?

There are a couple of points to be made here, actually. First is the fact that when used in this manner, the switch contacts need only handle what little base current is necessary to turn the transistor on, while the transistor itself handles the majority of the lamp's current. This may be an important advantage if the switch has a low current rating: a small switch may be used to control a relatively high-current load. Perhaps more importantly, though, is the fact that the current-controlling behavior of the transistor enables us to use something completely different to turn the lamp on or off.

Consider this example (Fig 4.11(a)), where a solar cell is used to control the transistor, which in turn controls the lamp.

Or, we could use a thermocouple (Fig 4.11(b)) to provide the necessary base current to turn the transistor on.

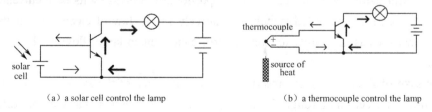

(a) a solar cell control the lamp   (b) a thermocouple control the lamp

Fig 4.11   examples of controlling the lamp

Even a microphone of sufficient voltage and current output could be used to turn the transistor on (Fig 4.12), provided its output is rectified from AC to DC so that the emitter-base PN junction within the transistor will always be forward-biased.

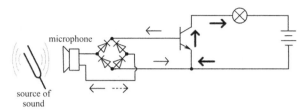

Fig 4.12　sound control the lamp

The point should be quite apparent by now: any sufficient source of DC current may be used to turn the transistor on, and that source of current need only be a fraction of the amount of current needed to energize the lamp. Here we see the transistor functioning not only as a switch, but as a true amplifier: using a relatively low-power signal to control a relatively large amount of power. Please note that the actual power for lighting up the lamp comes from the battery to the right of the schematic. It is not as though the small signal current from the solar cell, thermocouple, or microphone is being magically transformed into a greater amount of power. Rather, those small power sources are simply controlling the battery's power to light up the lamp.

Review:

Transistors may be used as switching elements to control DC power to a load. The switched (controlled) current goes between emitter and collector, while the controlling current goes between emitter and base.

When a transistor has zero current through it, it is said to be in a state of cutoff (fully nonconducting).

When a transistor has maximum current through it, it is said to be in a state of saturation (fully conducting).

4.3　Knowledge about Translation
（翻译知识 4——非谓语动词 to V）

1. to V

to V 又称动词不定式，兼有名词、形容词、副词特点，也保留了动词性，可用做主语、宾语、表语、定语、状语。常用来表示具体的（特别是未来的）一次性动作。

Resistor R_F **is made** variable to be able **to adjust** sufficient feedback voltage **to cause** oscillation. 电阻 R_F 是可调的，可调节到有足够的反馈电压以引起**电路的**振荡。（is made 不译，"电路的"为增补词语。用了两个 to V 表示一次性的动作、未来的动作）

The electromotive force creates the electric pressure that causes the current **to flow through a conductor.** 电动势产生电压，这电压使电流流过一个导体。（to V 作宾语补足语）

It is very difficult **to measure the passing current in insulators**. 测量绝缘体中通过的电流是很困难的。（不定式短语作主语时，尤其是不定式短语比较长时，往往引入 it 作形式主语，而把不定式短语放在谓语动词的后面。）

To describe the motion one must introduce the concept of time. 为了描述运动，必须引入时间的概念。（不定式短语作状语）

英语中有些动词后面只能接 to V，这些动词主要有：apply（申请），ask（要求），demand（要求），hope（希望），expect（期望），arrange（安排），plan（计划），prepare（准备）等。

I arrange to meet him at eight o'clock. 我安排八点和他会面。

I expect to be back on Sunday. 我希望星期天回来。

2．to V 与 V-ing 的区别

在翻译中要特别注意 to V 与 V-ing 的区别，否则造成曲解。

如：Stop to **smoke**. 停下来抽一支烟（一次性动作）。

Stop **smoking**. 戒烟（终止这一经常性的动作）。

I forgot **to do** it. 我忘记做这件事了（事没有做）。

I forgot **doing** it. 我忘记做过这件事了（事已做了）。

4.4　Exercises

1. Put the Phrases into English

① PN 结
② 阳极（正极）
③ 外阻（外部电阻）
④ 三极管
⑤ 一个小电流控制一个大电流
⑥ 基极电流
⑦ 增强型 MOS 管
⑧ P 沟道
⑨ 集成电路
⑩ 电子和空穴

2. Put the Phrases into Chinese

① semiconductor material
② forward biased
③ depend on the external circuit resistance
④ excessive reverse-biased voltage
⑤ is directly proportional to the amount of base current
⑥ may even appear almost as a short
⑦ cause stability problems for a transistor circuit
⑧ digital technology
⑨ the most popular technology
⑩ use two complementary types of transistors N-channel and P-channel

3. Sentence Translation

① In this condition the internal resistance of the diode is low and a large current will flow through the diode (depending on the external circuit resistance).

② Most transistors used today are NPN because this is the easiest type to make from silicon.

③ The acronym MOS stands for metal-oxide semiconductor, which historically denoted the gate, insulator, and channel region materials respectively.

④ The capacitor values depend on which circuit you use and what signal frequency you are amplifying.

⑤ If you really want "quick and dirty", this one transistor circuit is an "oldie but goodie".

⑥ Of course, it may seem pointless to use a transistor in this capacity to control the lamp.

⑦ This may be an important advantage if the switch has a low current rating: a small switch may be used to control a relatively high-current load.

⑧ When a transistor has zero current through it, it is said to be in a state of cutoff (fully nonconducting).

⑨ Please note that the actual power for lighting up the lamp comes from the battery to the right of the schematic.

⑩ The point should be quite apparent by now: any sufficient source of DC current may be used to turn the transistor on, and that source of current need only be a fraction of the amount of current needed to energize the lamp.

4. Translation (Pay More Attention to Black Words)

① In this case, the internal resistance is extremely high, **resulting** in very little current flow (**depending** on the diode's leakage current).

② **To turn** on an NPN bipolar transistor, the base must be more positive than the emitter (about +0.6 V).

③ Presently, the most popular technology for **realizing** microcircuits makes use of MOS transistors.

④ Both circuits need capacitors on the input and output **to block** DC while **passing** AC.

⑤ It is possible **to run** them from a single voltage by **using** two equal value resistors (R_b) **to create** a separate DC **grounding** point midway between U_{CC} and actual ground.

4.5　课文参考译文

4.5.1　半导体二极管

一个半导体二极管（简称二极管）可能是最简单的半导体器件。二极管是由半导体材料制成的 PN 结构成的。P 型材料（端）称为正极，而 N 型材料（端）称为负极（如图 4.1 所示）。

当二极管的正极电位高于负极电位（其差值大于开启电压，对锗管近似为 0.3 V，对硅管近似为 0.7 V）时称二极管是正向偏置，这时二极管的内部电阻是很小的，有一个较大的电流流过二极管，流过电流的大小取决于外部电路的电阻值。

当二极管的正极电位低于负极电位时称二极管反向偏置，这时二极管的内部电阻非常高，所以一个理想二极管可以阻挡反向的电流而让正向的电流通过。

二极管有很多用途，例如一个用电池的电器通常串联一个二极管来保护电器，以防止电池反接。如果电池放反了，二极管就阻挡（电池）的反向电流，起到保护电器的作用。

一个二极管的实际特性曲线并不是十分理想的，如图 4.2 所示。当理想二极管反向偏置时，电流不能通过，而实际二极管反向偏置时却有约 10 μA 的电流流过，（虽然很小，但仍不够理想)。如果加上足够大的反向电压,(PN)结会被击穿,让电流(反向)通过。选择二极管时一般要使其反向击穿电压远大于电路中可能出现的电压,二极管就不会击穿。

当二极管正向偏置时,只要很小的电压就可以使它导通。对硅管来说,正偏电压约为 0.7 V。虽然二极管正向偏置时,流过的电流可以比较大,但无论正向偏置还是反向偏置,电流过分大时都会损坏二极管。

图 4.3 给出一些二极管的照片。

4.5.2 NPN 双极型晶体管

有两类标准的双极型晶体管(通常中文中称为三极管):NPN 和 PNP 型,用不同的电路符号表示(如图 4.4 所示)。字母 N、P 表示制作三极管各层的材料,现在用的三极管大多数是 NPN 型,因为这种类型的硅管比较容易制作。

NPN 三极管由一块 N 型发射极(E)、一块 P 型基极(B)和一块 N 型集电极(C)组成。

一个晶体管可用于构成一个放大器。图 4.5 所示的电路用两个标准的 5 mm 的红色发光二极管和一个通用低功率 NPN 三极管构成(可以看到三极管中流过两个电流)。

当开关合上时,一个小电流流入三极管的基极,这个电流恰好使 B 极的发光二极管 B 微微发光,三极管放大这个小电流,使一个较大的电流通过 C 极流到 E 极。这个集电极电流很大,使发光二极管 C 很亮。

集电极电流 I_C 与基极电流 I_B 成正比,小于发射极电流 I_E,因为要使三极管导通必须有一个小的基极电流流入(发射极),三个电流之间的关系是 $I_E=I_C+I_B$。I_C 与 I_B 的比值称为三极管的电流放大系数,用来表示三极管的放大电流的能力,这个电流放大系数称为 β,当 C、E 两端的电压(U_{CE})保持不变时有 $\beta = \Delta I_C / \Delta I_B$。

要使一个 NPN 型的双极型晶体管导通,基极必须加略高于发射极的正向电压(对硅管来说约为+0.6 V)。当晶体管逐渐趋于饱和时,这个电压大约是+0.7 V,这时 C 极与 E 极之间的电阻很小,甚至几乎可以看成是短路。

当图 4.5 中的开关断开时,没有基极电流流过,所以三极管切断集电极电流,两个发光二极管都不亮(三极管截止),此时 C 极与 E 极之间的电阻很大,可以看成是开路(不通)。实际上三极管的性能没这么理想,这时仍有一个极小的漏电流(I_{CBO})从集电极流到基极,这个漏电流可能引起一个晶体管电路的稳定性(不好)的问题。

图 4.6 给出各种晶体三极管的照片。

4.5.3 MOS 晶体管

目前,实现(设计)微电路中最普遍的技术是利用 MOS 晶体管(简称 MOS 管)。缩写词 MOS 表示 M(金属)O(氧化物)S(半导体),分别表示(以前)用金属作门极,氧化物作绝缘层,半导体作沟道、基底等,根据载流子的不同,MOS 管可以分成两类:N 沟道和 P 沟道,N 沟道 MOS 管用电子传导电流,而 P 沟道 MOS 管用空穴传导电流。此外,N 沟道 MOS 管用一个正的门电压导通而 P 沟道 MOS 管用负的门电压导通。

MOS 管的功能是用小电压控制大电流。

根据电压条件的不同,MOS 又可分成增强型和耗尽型两种。增强型 MOS 晶体管和耗尽型 MOS 管的符号如图 4.7 所示。MOS 管实际上是四极的器件,第四极是与基底相连的。对数字电路来说,N 型沟道 MOS 管的基底总是与集成电路的最负端(即接地或 U_{SS})。同样,P

型沟道 MOS 管的基底总是与集成电路的最高电位端即 U_{DD} 相连，通常都这样连接，除非另外给出说明，所以（在电路图上）一般不标出基底的连接。

三极管技术主要应用单一管（全部用 NPN 型或全部用 PNP 型）形式（设计电路），与之不同的是 MOS 管设计的电路一般用两种互补型的晶体管，同时含有 N 沟道 MOS 管和 P 沟道 MOS 管的互补的 MOS 管集成电路称为 CMOS 电路。

4.5.4 理想运算放大器

在电子电路的学习中基本且重要的电路（器件）是理想电压放大器或称理想运算放大器。一般来说，理想运算放大器有两个输入端：v_1 与 v_2，和一个输出端 v_o。输入和输出之间的关系为 $v_o=A(v_1-v_2)$，其中 A 称为放大器的增益。注意因为理想运算放大器的输入阻抗 $R=\infty$，当这个放大器连接到任何电路中，都没有电流流入（运放的）输入端（实际上这个电流很小，所以这里是忽略不计了）。另外，因为输出电压 v_o 是一个理想电压源（模型）两端的电压，所以不管输出端接什么器件，都有 $v_o=A(v_1-v_2)$（即忽略了输出电阻），即增益为 A 的理想运算放大器可简单地用图 4.8 表示，输入端标为 "−" 表示该端输入电压与输出电压反相，称为反相输入端，输入端标为 "+" 则表示该端输入电压与输出电压同相，称为同相输入端。

4.6 阅读材料参考译文

4.6.1 音频放大电路

对那些喜欢做音频电路实验但不想去计算偏置电阻的人，我们给出两个简单的放大器电路（都用 9 V 电压源），其中一个用集成运算放大器构成（图 4.9(a)），另一个用一个三极管构成（图 4.9(b)）。

两个电路的输入和输出端都需要电容来隔直流通交流。电容值与你所选取的电路有关，并与你要放大的信号频率有关，至少为 1 μF，低频信号可能需要比较大的电容值。

集成运放一般要两个电压源供电（+U 和−U），如果想只用一个电压源供电需用两个等值电阻 R_b 接在电压源和实际接地点中间，创建一个单独的直流接地点，用来对集成运放提供偏置电压。直流接地点通过一个旁路电容与实际接地点相连。R_b 的值并不重要，10 kΩ 就可以了，为了减小输出的直流偏差，R_b 应取精度为 1% 的电阻。

放大电路的增益是通过 R_1 和 R_2 来设置的（电压增益 $A_U = R_2/R_1$），R_2 应该至少取 2 kΩ 或再大些的电阻，使集成运放的负载不至于太大。如果你用一个 TTL 器件如（老器件）741（一种集成运放型号），其输出电压最低为 2 V，最高则比 U_{CC} 低 2 V。所以对一个 9 V 电池供电来说，输出电压的最大变化范围只能是 5 V，从 2~7 V。如果你想要输出电压的范围大到从 U_{CC} 到地，则可以用 CMOS 器件如 CA3130，所加的 U_{CC} 可以在运放允许的范围内取最高值。CA3130 需要一个 100pF 的补偿电容。

如果你想要简单方便的放大电路，这款三极管电路虽然比较老，效果却不错（如图 4.9(b) 所示），图中把偏置电阻 R_b 直接连接到集电极的接法，有两个好处，一是这种偏置不可能引起三极管饱和或截止，二是这个电路构成的负反馈可以减小电路中的干扰。这个电路可能不如集成运放电路性能好，但它是一个有用的电路。至于这个电路的增益，你必须用不同的 R_c 值去试，并使 $R_b=100R_c$，通过测量来获取你需要的增益。（这个电路的）输入信号峰-峰值电压不能超过 15 mV。

4.6.2 三极管用做开关

因为一个三极管的集电极电流是正比于其基极电流的，所以可以把三极管作为一种电流控制开关。通过输入一个相当小的基极电流，可以起到控制一个较大的集电极电流的作用。

为了便于说明，我们通过用一个三极管取代开关来说明三极管如何控制流过电灯的电流（如图 4.10 所示）。记住通过三极管的受控电流是从集电极流到发射极。因为这个流过灯泡的电流是我们想要控制的，所以我们必须把三极管的集电极和发射极放在开关的两个接触端，我们还必须保证灯泡的电流的流向与发射极箭头的方向一致，以保证三极管的 PN 结偏置是正确的。

这个例子选用 NPN 型三极管构成。

要有基极电流，我们还要加一条电路，因为三极管的基极不接，基极电流为零，这个三极管就不能导通，灯也不亮。

如果基极开关断开，三极管的基极悬空（即不与任何地方相连接）没有电流流过。这种状态称三极管截止（如图 4.10(a)所示）。如果基极开关合上（如图 4.10(b)所示），则电子将从三极管的发射极流到基极，通过开关，流过左边的灯泡，回到电池的正极（注：电子的流向与电流方向相反）。这个基极电流将使更大量的电子从发射极流到集电极，点亮灯泡，这时电路电流达到最大值，这种状态称三极管饱和。

当然这样利用三极管去控制一个灯看起来毫无意义，毕竟只要用一个开关就可以控制一个灯了，不是吗？如果这个电路仍用一个开关去控制灯，如果仅是间接地用开关控制灯，三极管控制电流的意义是什么呢？为什么不干脆就用开关直接控制灯的电流？

实际上这里有两点：首先是采用这种方法控制时，开关中仅流过很小的基极电流使三极管导通，而三极管（集电极）流过的是灯中的大电流。这是一个优点：如果开关只允许流过小电流，就可以用一个小（额定电流）开关去控制一个相当大电流的负载。另点可能更重要，就是三极管的电流控制特性可以使我们用其他各种方法来控制灯。

可以利用一个太阳能电池来控制三极管从而控制灯（如图 4.11(a)所示）。

或者可以利用一个热电耦来提供基极电流使三极管导通（用温度控制灯）（如图 4.11(b)所示）。

如果通过整流，把话筒输出的交流信号变成直流信号，使三极管的发射极-基极的 PN 结正偏，甚至一个话筒的输出电压和电流可用来使三极管导通(如图 4.12 所示)（用声音控制灯）。

现在意义应该很清楚了，一个小小的直流电流源可以用来使三极管导通，直流电源的电流与灯泡中所流过电流相比是很小的。这里我们看到三极管的作用不但是一个开关，而且是一个放大器：用一个相当小功率的信号去控制一个相当大功率（的电器）。请注意实际点亮灯的能量来自于示意图右侧的电池，并不是什么来自太阳能电池、热电耦或话筒的信号被神奇地放大了（转换成一个大能量的信号），这些小信号源只是简单地起一个控制电池能源点亮灯泡的作用。

小结：

三极管可用做控制直流电源连接到负载的开关，当控制电流流过发射极和基极时，被控制电流流过发射极和集电极。

当三极管中没有电流流过时，称三极管处于截止状态（非导通）。

当三极管中流过的电流达到最大值时，称三极管处于饱和状态（完全导通）。

Unit 5 Power Supplies

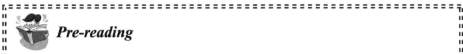

Pre-reading

Read the following passage, paying attention to the question.
1) What is the function of an electronic power supply?
2) What does an electronic power supply include ?

5.1 Text

5.1.1 Information on Power Supplies

There are many types of electrical and electronic power supplies, providing various alternating current (AC) and direct current (DC) voltages for equipment operation. DC voltages are needed for most electronic circuits, and generally an electronic power supply is considered as a device that converts AC into DC.

The AC voltage from the power lines can be rectified directly or passed through an isolation transformer (a turn's ratio of 1:1). Depending on the value of DC voltage needed, the transformer may be of either a step-up or step-down type. After the transformer, the AC is rectified into pulsating DC by diodes in the form of a half-wave rectifier, a full-wave rectifier, or a bridge (full-wave) rectifier. The pulsating DC is then filtered or smoothed out by capacitors, inductors, and resistors so as to produce a constant DC output.

5.1.2 Bridge (Full-wave) Rectifier

The bridge rectifier produces a full-wave DC output with the use of four diodes (Fig 5.1). When the positive half-cycle appears, the anode of diode 1 is positive and it conducts, and the cathode of diode 2 is negative and it conducts. Current flows from the AC source through diode 1, through R_L, through diode 2, and back to the source. Diodes 3 and 4 are reversed biased and no current flows through them. They appear as open circuits. On the negative half-cycle the anode of diode 4 is positive and it conducts, and the cathode of diode 3 is negative and it conducts. Current flows from the AC source through diode 4, through R_L, through diode 3, and back to the source. Diodes 1 and 2 are reverse. Voltage output U_o is the peak voltage less two times the U_F across the diodes (two diodes are conducting at a time):

$$U_o = U_{peak} - 2U_F$$

Full-wave bridge rectifier modules containing four diodes in a single unit are commercially available.

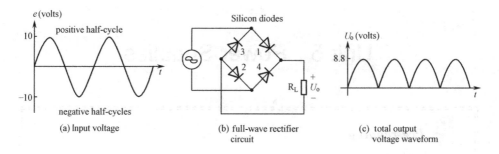

(a) Input voltage (b) full-wave rectifier circuit (c) total output voltage waveform

Fig 5.1 full-ware rectifier

5.1.3 Filter

Capacitors are used in power supplies to smooth out pulsating DC and to produce steady DC voltage. A simple capacitor filter is connected across the output. When this peak voltage begins to decrease, the stored electrons in the negative side of the capacitor will discharge and flow through the load in an attempt to keep the voltage constant across the capacitor. Another pulsating DC comes soon after the discharge and charges the capacitor to peak voltage again. The slight charging and discharging of the capacitor produces a ripple voltage (AC component) that is superimposed on the top of the steady DC (Fig 5.2).

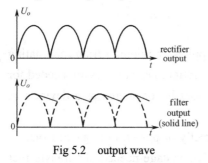

Fig 5.2 output wave

5.1.4 Zener Diode

A Zener diode is a specially doped diode that operates the same as a regular diode in the forward-biased condition (it allows large current to flow). However, in the reverse-biased condition (Fig 5.3), it will not conduct until it reaches the zener voltage (U_z) at which it is designed. At this point the Zener diode conducts current in the reverse direction, while maintaining the zener voltage across its terminals (Fig 5.4). The amount of current that flows through it is determined by two factors, a series (current-limiting) resistor (R_S) and the parallel load resistance (R_L).

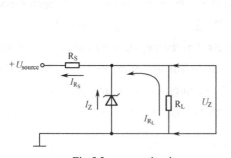

Fig 5.3 zener circuit

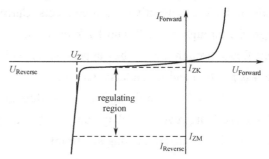

Fig 5.4 the output characteristic of a zener

Resistor R_S is found by the formula $R_S = U_{R_S} / I_z$ where $U_{R_S} = U_{source} - U_z$. With no load, a specific amount of current ($I_z = I_{R_S}$) flows through the Zener diode and R_S. Voltage drop U_{R_S} plus

U_z will equal U_{source} (U_{source} should be at least 1 V greater than U_z). When a load is connected across the Zener diode, the diode current decreases by the amount drawn by the load so that the current through R_S remains constant ($I_z = I_{R_S} - I_{R_L}$). The Zener diode maintains the voltage across its terminals by varying the current that flows through it.

5.1.5 Linear Voltage Regulator

The 78×× (also sometimes known as LM78××) series of devices is a family of self-contained fixed linear voltage regulator integrated circuits (Fig 5.5). The 78×× family is a very popular choice for many electronic circuits which require a regulated power supply, due to their ease of use and relative cheapness. When specifying individual ICs within this family, the ×× is replaced with a two-digit number, which indicates the output voltage the particular device is designed to provide (for example, the 7805 has a 5 volt output, while the 7812 produces 12 volts).

Fig 5.5 78×× ICs

The 78×× are positive voltage regulators, meaning that they are designed to produce a voltage that is positive relative to a common ground. There is a related line of 79×× devices which are complementary negative voltage regulators. 78×× and 79×× ICs can be used in combination to provide both positive and negative supply voltages in the same circuit, if necessary.

Technical Words and Phases

alternate	[ˈɔltəːneɪt]	*adj.* 交替的，轮流的，预备的 *v.* 交替，轮流，改变
discharge	[dɪsˈtʃɑːdʒ]	*n.* 放电器 *vt.* 放（电）
filter	[ˈfɪltə(r)]	*n.* 滤波器，过滤器，滤光器 *vt.* 过滤
isolation	[aɪsəˈleɪʃ(ə)n]	*n.* 隔绝，孤立，隔离，绝缘
pulsate	[pʌlˈseɪt]	*vi.* （脉等）搏动，脉动[冲]，波动
rectify	[ˈrektɪfaɪ]	*v.* 整流，检波
ripple	[ˈrɪp(ə)l]	*n.* 脉动，波动；波纹（波动） *vt.* 使……波动
smooth	[smuːð]	*adj.* 平滑的 *vt.* 使光滑，使优雅，消除 *vi.* 变平滑，变平静
transformer	[trænsˈfɔːmə(r)]	*n.* 变压器（变换器，互感器）

electronic power supply	电子稳压电源
bridge(full-wave)rectifier	桥式（全波）整流器
two diodes are conducting at a time	两个二极管同时导通
Zener diode	齐纳二极管，稳压管
commercially available	有商品供应的（即可以直接买到的）

5.2 Reading Materials

5.2.1 about the IEEE (Institute of Electrical and Electronics Engineers)

21 January 2009 - IEEE is commemorating its 125th Anniversary this year with a variety of activities surrounding the theme of "Celebrating 125 Years of Engineering the Future."

The IEEE (Eye-triple-E) is a non-profit, technical professional association of more than 375 000 members including nearly 80 000 student members in more than 160 countries. The full name is the Institute of Electrical and Electronics Engineers, although the organization is most popularly known and referred to by the letters I-E-E-E.

Through its members, the IEEE is a leading authority in technical areas ranging from computer engineering, biomedical technology and telecommunications, to electric power, aerospace and consumer electronics, among others.

Through its technical publishing, conferences and consensus-based standards activities, the IEEE produces nearly a third of the world's technical literature in electrical engineering, computer science and electronics. Holds annually more than 400 major conferences and has nearly 900 active standards and more than 400 standards in development.

The benefits of IEEE membership include these offerings:

- Membership in one or more of 38 IEEE societies and four technical councils spanning the range of electrotechnologies and information technologies.
- More than 300 local organizations worldwide for member networking and information sharing.
- Educational opportunities to ensure engineers' technical vitality.
- More than 1 600 student branches at universities worldwide.
- Special cost-saving and value-added benefits for MEMBERS ONLY.
- Prestigious awards and recognition of technical and professional achievements.
- Opportunities for volunteering, leadership and participation in a variety of IEEE activities.

（本文摘自于 IEEE 网站）

5.2.2 Robots

The vast majority of robots do have several qualities in common.

First of all, almost all robots have a movable body. Some only have motorized wheels, and others have dozens of movable segments, typically made of metal or plastic. Like the bones in your body, the individual segments are connected together with joints.

Robots spin wheels and pivot jointed segments with some sort of actuator. Some robots use electric motors and solenoids as actuators; some use a hydraulic system; and some use a pneumatic system (a system driven by compressed gases).

A robot needs a power source to drive these actuators. Most robots either have a battery or they plug into the wall. Hydraulic robots also need a pump to pressurize the hydraulic fluid, and pneumatic robots need an air compressor or compressed air tanks.

The actuators are all wired to an electrical circuit. The circuit powers electrical motors and solenoids directly, and it activates the hydraulic system by manipulating electrical valves.

Not all robots have sensory systems, and few have the ability to see, hear, smell or taste. The most common robotic sense is the sense of movement - the robot's ability to monitor its own motion.

The robot's computer controls everything attached to the circuit. To move the robot, for example, the computer switches on all the necessary motors and valves. Most robots are reprogrammable - to change the robot's behavior, you simply write a new program to its computer.

A robotic hand (Fig 5.6), developed by NASA (National Aeronautics and Space Administration), is made up of metal segments moved by tiny motors. The hand is one of the most difficult structures to replicate in robotics.

The most common manufacturing robot is the robotic arm (Fig 5.7). A typical robotic arm is made up of seven metal segments, joined by six joints. The computer controls the robot by rotating individual step motors connected to each joint (some larger arms use hydraulics or pneumatics). Unlike ordinary motors, step motors move in exact increments. This allows the computer to move the arm very precisely, repeating exactly the same movement over and over again. The robot uses motion sensors to make sure it moves just the right amount.

Some mobile robots are controlled by remote - a human tells them what to do and when to do it. The remote control might communicate with the robot through an attached wire, or using radio or infrared signals. Remote robots, often called puppet robots, are useful for exploring dangerous or inaccessible environments, such as the deep sea or inside a volcano.

Some robots can interact socially. Kismet, a robot at M.I.T's Artificial Intelligence Lab (Fig 5.8), recognizes human body language and voice inflection and responds appropriately. Kismet's creators are interested in how humans and babies interact, based only on tone of speech and visual cue. This low-level interaction could be the foundation of a human-like learning system.

Fig 5.6   robotic hand   Fig 5.7   robotic arm   Fig 5.8   the humanoid robot

5.2.3 How Power Grids Work

Electrical power is a little bit like the air you breathe: You don't really think about it until it is missing. Power is just "there", meeting your every need, constantly. It is only during a power failure, when you walk into a dark room and instinctively hit the useless light switch that you realize how important power is in your daily life. You use it for heating, cooling, cooking, refrigeration, light, sound, computation, entertainment. Without it, life can get somewhat cumbersome.

Power travels from the power plant to your house (Fig 5.9) through an amazing system called the power distribution grid. The grid is quite public – if you live in a suburban or rural area, it is right out in the open and chances are for all to see. Your brain likely ignores all of the power lines because it has seen them so often. In this article, we will look at all of the equipment that brings electrical power to your home.

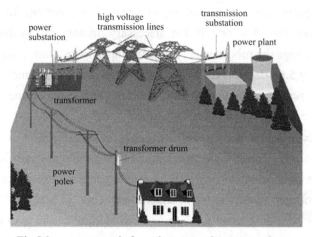

Fig 5.9 power travels from the power plant to your house

1. the Power Plant

Electrical power starts at the power plant. In almost all cases, the power plant consists of a spinning electrical generator. Something has to spin that generator – it might be a water wheel in a hydroelectric dam, a large diesel engine or a gas turbine. But in most cases, the thing spinning the generator is a steam turbine. The steam might be created by burning coal, oil or natural gas, or the steam may come from a nuclear reactor.

No matter what it is that spins the generator, commercial electrical generators of any size generate what is called 3-phase AC power. To understand 3-phase AC power, it is helpful to understand single-phase power first.

2. Alternating Current

Single-phase power is what you have in your house. You generally talk about household electrical service as single-phase, 220-volt AC service. If you use an oscilloscope and look at the power found at a normal wall-plate outlet in your house, what you will find is that the power at the wall plate looks like a sine wave, and that wave oscillates between -311 volts and 311 volts (the peaks are indeed at 311 volts; it is the effective (rms) voltage that is 220 volts). The rate of oscillation for the sine wave is 50 cycles per second. Oscillating power like this is generally referred to as AC, or alternating current.

3. the Power Plant Produces AC

The power plant produces three different phases of AC power simultaneously, and the three phases are offset 120 degrees from each other. There are four wires coming out of every power plant: the three phases plus a neutral or ground common to all three. If you were to look at the three phases on a graph, they would look like this (Fig 5.10) relative to ground.

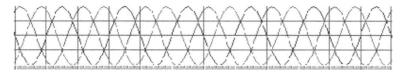

Fig 5.10 the waveform of 3-phase AC

4. Transmission Substation

The three-phase power leaves the generator and enters a transmission substation at the power plant. This substation uses large transformers to convert the generator's voltage (which is at the thousands of volts level) up to extremely high voltages for long-distance transmission on the transmission grid.

You can see at the back several three-wire towers leaving the substation. Typical voltages for long distance transmission are in the range of 155 000 to 765 000 volts in order to reduce line losses. A typical maximum transmission distance is about 483 km. High-voltage transmission lines are quite obvious when you see them. They are normally made of huge steel towers (see in Fig 5.9).

5. the Distribution Grid

For power to be useful in a home or business, it comes off the transmission grid and is stepped-down to the distribution grid. This may happen in several phases. The place where the conversion from "transmission" to "distribution" occurs is in a power substation. A power substation typically does two or three things:

It has transformers that step transmission voltages (in the tens or hundreds of thousands of volts range) down to distribution voltages (typically less than 10 000 volts).

It has a "bus" that can split the distribution power off in multiple directions.

It often has circuit breakers and switches so that the substation can be disconnected from the transmission grid or separate distribution lines can be disconnected from the substation when necessary.

6. Distribution Bus

The power goes from the transformer to the distribution bus: In this case, the bus distributes power to two separate sets of distribution lines at two different voltages. The smaller transformers attached to the bus are stepping the power down to standard line voltage (7 200 V) for one set of lines, while power leaves in the other direction at the higher voltage of the main transformer. The power leaves this substation in two sets of three wires, each headed down the road in a different direction.

The three wires at the top of the poles are the three wires for the 3-phase power. The fourth wire lower on the poles is the ground wire. In some cases there will be additional wires, typically phone or cable TV lines riding on the same poles.

7. Taps and at the house

A house needs only one of the three phases, so typically you will see three wires running down a main road, and taps for one or two of the phases running off on side streets.

Past a typical house runs a set of poles with one phase of power (at 7 200 volts) and a ground wire (although sometimes there will be two or three phases on the pole, depending on where the house is located in the distribution grid). At each house, there is a transformer drum (Fig 5.9) attached to the pole. The transformer's job is to reduce the 7 200 volts down to the 220 volts that makes up normal household electrical service. The 220 volts enters your house through a typical watt-hour meter.

5.3 Knowledge about Translation
（翻译知识 5——非谓语动词 V-ed）

动词的 V-ed 形式与 be 结合构成被动态，与 have (had)结合构成完成时，因此 V-ed 形式本身含有被动与完成的意思，它可在句子中担任定语、状语等，保留了动词性，表示这一动作是已完成的或所修饰名词所（被动）接受的。

Multidigit displays consist of two or more seven-segment displays **contained** in a single package or module. 多位数码显示器由两个或更多的七段显示器（被封装在一个单元中）组成。

1. V-ed 作定语

单个 V-ed 作定语一般放在名词前面（也可以放在后面），V-ed 短语作定语一般放在名词之后。在作定语时，分词在意思上接近一个定语从句。

The work **done** is the product of the force and the distance. 所做的功是力和距离的乘积。

Heat is the energy **produced by the movement of molecules**. 热是分子运动所产生的能量。

A semiconductor diode consists of a PN junction **made of semiconductor material**. 一个半导体二极管是由一个半导体材料制成的 PN 结构成的。

2．V-ed 作状语

V-ed 作状语可表示动作发生的背景和情况。

All metals are fairly hard, **compared to** nonmetals. 与非金属相比，金属是相当硬的。

Tonight, **lit** by countless electric lights, all the halls are as bright as day. 今晚无数个电灯把所有的大厅照耀得如同白昼。

3. V-ing 与 V-ed 的区别

正如前面已提到的，V-ing 通常有主动的意思，V-ed 通常有被动的意思，在阅读翻译时要注意这一点，才能区分动作的发出者和对象。试比较：

I heard someone **opening** the door. 我听见有人开门。

I heard the door **opened**. 我听见门给（谁）打开了。

The **switched (controlled)** current goes between emitter and collector, while the **controlling** current goes between emitter and base. 当控制电流流过发射极和基极时，被控制电流流过发射极和集电极。

5.4 Exercises

1. Put the Phrases into English

① 稳压电源
② 桥式整流器
③ 脉冲直流电
④ 二极管的正极
⑤ 峰值电压
⑥ 电容滤波器
⑦ 充电和放电
⑧ 稳压管
⑨ 电气电子工程师学会
⑩ 专业技术组织

2. Put the Phrases into Chinese

① equipment operation
② device that converts AC into DC
③ the power lines
④ depending on the value of DC voltage needed
⑤ a half-wave rectifier
⑥ so as to produce a constant DC output
⑦ in the negative side of the capacitor
⑧ flow through the load
⑨ in the forward-biased condition
⑩ a series (current-limiting) resistor

3. Sentence Translation

① However, in the reverse-biased condition, it will not conduct until it reaches the zener voltage (U_z).

② After the transformer, the AC is rectified into pulsating DC by diodes in the form of a bridge rectifier.

③ Before the capacitor discharges too far, another DC pulse arrives to recharge the capacitor back to the peak voltage.

④ The full name is the Institute of Electrical and Electronics Engineers Inc., although the organization is most popularly known and referred to by the letters I-E-E-E.

⑤ The IEEE produces 30 percent of the world's published literature in electrical engineering, computers and control technology.

⑥ It is only during a power failure, when you walk into a dark room and instinctively hit the useless light switch that you realize how important power is in your daily life.

⑦ Power travels from the power plant to your house through an amazing system called the power distribution grid.

⑧ You generally talk about household electrical service as single-phase, 220-volt AC service.

⑨ Typical voltages for long distance transmission are in the range of 155 000 to 765 000 volts in order to reduce line losses.

⑩ It often has circuit breakers and switches so that the substation can be disconnected from the transmission grid when necessary.

4. Read Industrial Robots, Choose the Best Answer for Each of the Following

① Which of the following is true?

   A. A robot is designed to have a brain.

   B. A robot is designed without a brain.

   C. A robot is controlled by a computer.

   D. A robot is designed not to interact with the real world.

② According to the passage, most robots can't

   A. Make intelligent decisions and talk with people

   B. Assemble components

   C. Perform inspections

   D. Grasp objects

③ Which of the following is not true?

   A. The most common manufacturing robot is the robotic arm.

   B. The hand is one of the most difficult structures to replicate in robotics.

   C. Remote robots are useful for exploring dangerous or inaccessible environments.

   D. Hydraulic robots don't need a pump to pressurize the hydraulic fluid.

④ What is not the robots' qualities in common?

   A. Almost all robots have a movable body .

   B. All robots have sensory systems.

   C. Robots spin wheels and pivot jointed segments with some sort of actuator.

   D. A robot needs a power source to drive his actuators.

⑤ Which of the following is not true according to the text?

   A. Hydraulic robots need a pump to pressurize the hydraulic fluid.

   B. The most common robotic sense is the sense of movement.

   C. Most robots are reprogrammable.

   D. Ordinary motor allows the computer to move the arm very precisely.

5.5 课文参考译文

5.5.1 关于（稳压）电源

有许多类型的电气和电子的供电设备为（工作）设备的运行提供各种各样的交流和直流电压（能量）。许多电子电路都需要直流电压源供电，通常一个电子稳压电源就是把交流电源转换成直流电源的器件（设备）。

来自（电力网）电源线的交流电压可以直接或通过一个绝缘变压器（其变压比为 1:1）后进行整流。根据所需要的直流电压值的不同，变压器还可以是一个升压变压器或降压变压器。从变压器副边流出的交流电流通过二极管构成的半波整流器、全波整流器或桥式整流器整流成为脉冲直流电。然后这脉冲直流电通过电容、电感和电阻构成的滤波电路滤波，从而产生较稳定的直流电压输出。

5.5.2 桥式（全波）整流器

用 4 个二极管构成的桥式整流器（如图 5.1 所示）产生一个全波直流电压输出。当交流电压源为正半周时，二极管 1 的正极为正，二极管 1 导通，二极管 2 的负极为负，二极管 2 也导通，从交流电源（正极）流出的电流流经二极管 1、电阻 R_L、二极管 2 再回到电压源（负极）。这时二极管 3、4 是反向偏置的，没有电流流过。就好像是断开的。在（交流电压源的）负半周，二极管 4 的正极为正，二极管 4 导通，二极管 3 的负极为负，二极管 3 也导通，从交流电源（正极）流出的电流流经二极管 4、电阻 R_L、二极管 3 再回到电压源（负极）。这时二极管 1、2 是反向偏置的。输出电压值 U_o 为峰值电压减去两倍的二极管正向导通电压。（每一次都有两个二极管导通）。

$$U_o = U_{peak} - 2U_F$$

目前可以直接买到含有 4 个二极管构成全波桥式整流器的模块器件。

5.5.3 滤波器

在稳压电路中用电容对脉冲直流电进行平滑（滤波），以产生稳定的直流电压。一个简单的电容滤波器是（并联）接在输出电路中的，当这个峰值电压开始减少时，在电容负极板储藏的电子开始放电，放电电流过负载，企图保持电容两极的电压不变。电容放电不久，另一个直流脉冲又到了，再次对电容充电使它达到峰值电压。电容的微小的充电和放电产生一个小波纹电压（交流分量），这个小波纹电压叠加在稳定直流电压上（如图 5.2 所示）。

5.5.4 齐纳二极管（稳压管）

稳压管是一种特殊的二极管，在正偏的条件下，它与一般的二极管有相同的特性（可以流过一个大电流）。但是，在反向偏置时（如图 5.3 所示），在外加电压低于稳压电压（U_Z）（设计时确定的）时它不导通，在外加电压等于稳压电压（U_Z）时稳压管反向导通，同时维持稳压管两端的电压为稳压值（如图 5.4 所示）。流过稳压管的电流的大小由两个因子决定：一个为串联的（限流）电阻（R_S），另一个为并联的负载电阻（R_L）。

电阻 R_S 由公式 $R_S = U_{R_S} / I_Z$ 确定，其中 $U_{R_S} = U_{source} - U_Z$，在没有负载时，一个特定大小的电流（$I_Z = I_{R_S}$）流过稳压管和 R_S，电压降 U_{R_S} 加 U_Z 等于 U_{source}（U_{source} 至少要比 U_Z 高 1 V）。当一个负载并连到稳压管，流过稳压管的电流由于负载的分流而减小，所以通过 R_S 的电流保持为常数（$I_Z = I_{R_S} - I_{R_L}$）。稳压管通过改变流过它的电流来维持稳压管两端的电压稳定。

5.5.5 线性稳压器

78××（有时称为 LM78××）是线性稳压集成电路系列器件（如图 5.5 所示），很多电子电路需要稳压时常选用 78×× 系列芯片，因为用起来方便且价格便宜。这个系列的集成电路器件有不同的（稳定）输出电压，对于具体的器件，×× 用两位数字取代，表示这个器件的输出电压值（例如 7805 表示输出电压为 5 V，7812 表示输出电压为 12 V）。

78×× 是正电源稳压器，即这类芯片的输出电压相对于其公共端是正的，还有一个相应的 79×× 系列是互补的负电源稳压器。需要的话，78×× 和 79×× 系列集成电路可以在同一电路中组合应用，提供正、负电源电压。

5.6 阅读材料参考译文

5.6.1 关于 IEEE（电气电子工程师学会）

2009 年 1 月 21 日起，IEEE 将在这一年围绕"庆祝最有前景的工程师协会成立 125 周年"这个主题举行一系列的活动，以庆祝其成立 125 周年。

IEEE（注意是 3 个 E）是一个有 160 个国家的多达 375 000 会员（其中包括近 80 000 学生会员）的非营利的专业技术学会，其全称是电气电子工程师学会，虽然这个组织众所周知的名字为 IEEE。

通过它的成员，IEEE 成为一个在计算机工程、生物医学技术、通信技术、电源、航空和日用电子产品等众多技术领域的指导性权威机构。

通过出版技术期刊，举办专业会议和制定统一的标准等活动，IEEE 在电气工程、计算机和控制技术方面的出版物几乎占世界出版的技术文献的 1/3，每年举行超过 400 多个专业学术会议，已制定了近 900 个正在实施的标准，另有 400 多个标准正在研究开发。

作为 IEEE 成员有如下的优势：

- 作为 38 个 IEEE 组织和 4 个涉及电子技术和通信技术理事会中的一个或多个组织的会员（即可以同时参加几个 IEEE 的组织）。
- 300 多个遍布世界的地方组织成员网络和信息共享。
- 为保证工程师的技术活力（而提供的）教育机会。
- 在世界各地大学中的 1 600 多个学生分支机构。
- 专为成员提供的节约成本和增值的优惠。
- 声望很高的技术和专业成就的奖励和奖金。
- 在 IEEE 的各种活动中为志愿者，领导者和参与者提供机会。

5.6.2 机器人

绝大部分机器人都有一些共同的性质。

首先，几乎所有的机器人都有一个可移动的部件。有些只有可动的轮子，有些则有很多可动的部件，一般用金属或塑料制成，就像人身体上的骨头，这些部件用关节连接在一起。

机器人用一些执行器使轮子和轴关节部件运动。有的机器人用电动机和电磁（螺线管）作为执行器，另一些则用液压系统和气压系统（用压缩空气驱动的系统）作为执行器。

机器人需要用电源来驱动这些执行器，很多机器人用电池或（墙上的）交流电源供电。液压的机器人也要用液压泵来给液体加压，气动机器人则要用一个空气压缩机或压缩空气罐（给气体加压）。

执行器都要与一个电路相连接，电路直接驱动电动机和电磁螺线管，或通过操纵电磁阀使液压系统动作。

并不是所有的机器人都是有感觉的，极少数（几乎没有）机器人可以看、听、闻或尝，大部分常用的机器人只有运动传感器——机器人可以监控它自身的动作。

机器人的计算机通过电路控制一切，例如，机器人移动时，计算机控制所有相应的电动机和电磁阀。大部分机器人是可编程的——要改变机器人的动作，只要把新的程序输给计算机。

由美国宇航局（或美国太空总署）开发的机器手（如图 5.6 所示）是由微型电动机控制的可移动的金属部件组成的。在机器人中人手是最难仿造的结构之一。

制造业中最常用的机器人是机器手臂（如图 5.7 所示）。典型的机器手臂由 7 个金属部件、6 个关节组成。计算机通过分别转动与每个关节相连接的步进电动机控制机器手臂（有些较大的手臂采用液压或气压系统）。与普通电动机不同的是，步进电动机可以精确控制移动量，使计算机可以很精确地控制手臂的动作，可以准确地控制同样的动作重复一次又一次。机器手臂用运动传感器确保其移动到位。

有些可移动的机器人是遥控的——人们告诉它做什么和什么时候做。遥控是通过导线、射频信号或红外信号与机器人通信。遥控机器人常称为木偶机器人，在探索危险环境或人不能到达的环境如海洋深处或火星内部时很有用。

有些机器人能与人交流，美国麻省理工学院人工智能实验室制作的机器人 Kismet（如图 5.8 所示），能识别人的肢体语言和声音变化，并作适当的响应，Kismet 的设计者从人与婴儿仅通过语调和动作进行交流中受到启发制作了 Kismet。这种低级的交流可能是人类学习系统的基础。

5.6.3 电力网是如何工作的

电力网有点像你呼吸的空气，除非没有了否则你不会想到它。电一直在"那里"满足着你的需要，只有当停电时，当你在黑暗的房间里走动本能地想去打开无用的开关时，你会发现电在你的日常生活中是多么重要。你用电来取暖、制冷、煮饭、冷藏食物、照明、收听音乐、计算和各种娱乐等。没有电，生活就会有许多不方便的地方。

电从发电厂传到你家（如图 5.9 所示）是通过一个很了不起的系统，这个系统称为电力（分布）网。这个电力网非常普遍，如果你生活在郊区或农村，电力网就在野外，随处可见。正因为经常见到，所以你根本就没注意到所有这些电力线的存在。这里，我们来看一下把电能从发电站送到你家的各种设备。

1. 电厂

电从电厂开始，几乎所有的电厂是由旋转式发电机和带动发电机旋转的设备组成的，这个设备在水力发电站可能是一个水轮（机），也可能是一个大的柴油机或一个汽轮机。但大多数情况下是蒸汽轮机拖动发电机旋转。这蒸汽是通过燃烧煤、石油或天然气，或者通过核反应堆产生（热量烧水从而产生）的。

无论是何种方式拖动发电机，各种型号的（商用）电力发电机产生的都是三相交流电。为了理解三相交流电，先看单相交流电。

2. 交流电

单相交流电源就是你在家所用的电源。通常提到的民用电就是单相、220 V 交流电源。如果你用示波器看你家墙上电源插座的输出电压，你会在示波器的显示屏上看到电源的波形像一个正弦波，这个正弦波在-311 V 和 311 V 之间振荡（峰值确实是 311 V，有效值电压是 220 V）。正弦波振荡的速率是每秒 50 次。这种振荡电源通常称为交流电。

3. 电厂产生交流电

电厂同时产生三相交流电，三个交流电的相位互差120°，每个电厂出来有四根电力线，三根相线加上一根中线或三相的公共接地线。三相交流电相对于地的电压波形如图5.10所示。

4. 输电变电站

从发电机出来的三相交流电进入电厂的输电变电站。变电站用大变压器把发电机的电压（大约几千伏）升压为特别高的电压，为在电力网上的长距离输出做准备。

你可从变电站的后面看到一些三线的（电力）塔通向远处。为了减少线路传输损耗，典型的长距离传输电压在 15.5 万伏～76.5 万伏，典型的传输距离大约是 483 km。高压传输线是很显眼的，通常用巨大的钢塔组成（如图 5.9 所示）。

5. 配电网

电力在商业和民用中都是不可缺少的，来自输电网的电逐级降压进入配电网，这个过程可能分成几个阶段。把电从输电网转换到配电网的地方是变配电站，一个变配电站主要做这样两三件事：

它有一个降压变压器把传输网的电压（在几万伏到几十万伏的范围）降为配电网的电压（通常小于 1 万伏）。

它有一组传输线把（配电网的）电力送到四面八方。

它通常有断路器和开关，所以在需要时变配电站可以断开输电网的高压电力线或断开配电网的低压电力线。

6. 配电总线

从变压器出来的电流入配电总线：在这种情况中，配电总线的电力按两种不同的电压分成两组配电线，对其中一组配电线，通过与总线相连的小变压器把电压降低到标准的线电压（7 200V），主变压器出来的较高的电压的电力线则朝另一个方向传送。电力线按三线一组分成两组，朝着不同的方向离开这个变电站。

在电线杆顶部的三根线是三根相线，略低一点的第四根线是地线，有时还有一些附加的线如电话线、有线电视线等安装在同一根电线杆上。

7. 支线及进屋

民用的只是三相电中的一相，所以通常你会看到沿着主干道的三相线和通向小街小巷的一相或二相支线。

过去典型的房子是用很多带有一根单相电力线（7 200 V）和一根地线通电的电线杆送电到户的（虽然根据房子在配电线路中的位置不同，有时电线杆上有二根或三根相线）。在每个房子前的电线杆上有一个圆形变压器（如图 5.9 所示），变压器把 7 200 V 电压降为 220 V 家用标准电压，电力线通过一个电度表后把 220 V 的交流电源送到你的家。

Unit 6 Linear Circuit Analysis

> **Pre-reading**
>
> Read the following passage, paying attention to the question.
> 1) What is Kirchhoff's Current Law?
> 2) What is Kirchhoff's Voltage Law?
> 3) Why the sinusoid is an extremely important function?

6.1 Text

6.1.1 Electric Circuit

Have you ever wondered what happens when you flip a switch to turn on a light, TV, vacuum cleaner or computer? In all of these cases, you are completing an electric circuit, allowing a current, or flow of electrons, through the wires.

Fig 6.1 shows a simple circuit of a flashlight with a battery at one end and a flashlight bulb at the other end. When the switch is off, a complete circuit will not exist, and there will be no current. When the switch is on, there will be a complete circuit and a flow of current resulting in the flashbulb emitting light.

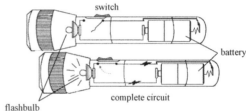

Fig 6.1   circuit of a flashlight

Circuits can be huge power systems transmitting megawatts of power over a thousand miles - or tiny microelectronic chips containing millions of transistors. This extraordinary shrinkage of electronic circuits made desktop computers possible. The new frontier promises to be nanoelectronic circuits with device sizes in the nanometers (one-billionth of a meter).

6.1.2 Ohm's Law

Suppose that some material is connected to the terminals of an ideal voltage source $u(t)$ as shown in Fig 6.2. Suppose that $u(t)=1$ V, then the electric potential at the top of the material is 1 V above the potential at the bottom. Since an electron has a negative charge, electrons in the material will tend to flow from bottom to top. Therefore, we say that a current tends to go from top to bottom through the material. Hence, for the given polarity, when $u(t)$ is a positive number, $i(t)$ will be a positive number with the direction indicated in Fig 6.2. If $u(t)=2$ V, again the potential at the top is greater 2 V than at the bottom, so

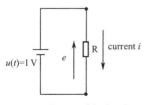

Fig 6.2   connect R to an ideal voltage source

$i(t)$ will again be positive. However, because the potential is now twice as large as before, the current $i(t)$ be greater (If the material is a "linear" element, the current will be twice as great). If the resulting current $i(t)$ is always directly proportional to the voltage for any function $u(t)$, the material is called a linear resistor.

To a linear resistor, we have:
$$R = \frac{u(t)}{i(t)} \quad \text{and} \quad i(t) = \frac{u(t)}{R}$$

The unit of resistance (volts per ampere) is referred to as Ohms and is denoted by the capital Greek letter omega Ω. Both equations above are also referred to as Ohm's law.

6.1.3 Kirchhoff's Law

It is a consequence of the work of the German physicist (1824—1887) that enables us to analyze an interconnection of any number of elements (voltage sources, current sources, and resistors), as well as other electronic devices. We will refer to any such interconnection as a circuit or a network.

We now present the first of Kerchief's two laws, his current law (KCL), which is essentially the law of conservation of electric charge:

For a given circuit, a connection of two or more elements shall be called a node (Fig 6.3). At any node of a circuit, at every instant of time, the sum of the currents into the node is equal to the sum of the currents out of the node.

$$\sum i_{in} = \sum i_{out}$$

An alternative, but equivalent, form of KCL can be obtained by considering currents directed into a node to be positive in sense and currents directed out of a node to be negative in sense, under this circumstance, the alternative form of KCL can be stated as follows:

$$\sum i = 0$$

At any node of a circuit, the currents algebraically sum to zero.

We now present the second of Kirchhoff's laws- the voltage law. To do this, we must introduce the concept of a "loop". Starting at any node in a circuit, we form a loop by traversing through elements and returning to the starting node, and never encountering any other node more than once (Fig 6.4). Kirchhoff's voltage law (KVL) is:

$$\sum u_+ = \sum u_-$$

In traversing any loop in any circuit, at every instant of time, the sum of the voltages having one polarity equals the sum of the voltages having the opposite polarity.

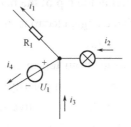
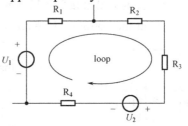

Fig 6.3 node of a circuit　　　　　　　　　Fig 6.4 loop of a circuit

An alternative statement of KVL can be obtained by considering voltages across elements that are traversed from plus to minus to be positive in sense and voltages across elements that are traversed from minus to plus to be negative in sense (or vice versa). Under this circumstance, KVL has the following alternative form.

$$\sum u = 0$$

Around any loop in a circuit, the voltages algebraically sum to zero.

6.1.4 Circuit Analysis Techniques

The process by which we determine a variable (either voltage or current) of a circuit is called analysis. Though: some simple circuits can be quite useful. We have already come across some simple circuits whose analysis was accomplished by applying the basic principles converted to data-Ohm's law, KCL and KVL. Nonetheless, we must also be able to analyze more complicated circuits-those in which it is simply not possible to conveniently write and solve a set of linear algebraic equations.

There are several distinct approaches that we can take. In one we write a set of simultaneous equations in which the variables are voltage, this is known as nodal analysis. In another we write a set of simultaneous equations in which the variables are currents, this is known as mesh analysis. Although nodal analysis can be used to all circuits, this is not the case for mesh analysis. A class of circuits known as "nonplanar" networks cannot be handled with mesh analysis. However, a similar approach - loop analysis - can be used. Here the variables are also current.

6.1.5 Sinusoidal Circuits

Step and impulse functions are useful in determining the responses of circuits when they are first turned on or when sudden or irregular changes occur in the input. This is called transient analysis. However, to see how a circuit responds to a regular or repetitive input-the steady-state analysis - function that is by far the most useful is the sinusoid.

The sinusoid is an extremely important and ubiquitous function. To begin with the shape of ordinary household voltage is sinusoidal, consumer radio transmissions are either amplitude modulation (AM), in which the amplitude of a sinusoid is changed or modulated according some information signal, or frequency modulation (FM), in which the frequency of a sinusoid is modulated.

We have following conclusions about the sinusoid:

① If the input of a linear, time-invariant circuit is a sinusoid, then the response is sinusoid of the same frequency.

② Finding the magnitude and phase angle of a sinusoidal steady-state response can be accomplished with either real or complex sinusoids.

③ If the output of a sinusoidal circuit reaches its peak before the input, the circuit is a lead network. Conversely, it is a lag network.

④ Using the concepts of phasor and impedance, sinusoidal circuits can be analyzed in the frequency domain in a manner analogous to resistive circuits by using the phasor versions of

KCL, KVL, nodal analysis, mesh analysis and loop analysis

Technical Words and Phases

admittance	[əd'mɪtəns]	*n.*	导纳（即电阻的倒数）
algebraically	[ældʒɪ'breɪɪkəlɪ]	*adv.*	代数地，用代数方法
consequence	['kɔnsɪkwəns]	*n.*	结果，推理，推论，因果关系，重要的地位
depict	[dɪ'pɪkt]	*vt.*	描述
impedance	[ɪm'pɪːdəns]	*n.*	阻抗（包括电阻与电抗）
loop	[luːp]	*n.*	回路
megawatt	['megəwɔt]	*n.*	兆瓦，百万瓦特（电能计量单位）
mesh	[meʃ]	*n.*	网络
node	[nəud]	*n.*	节点
nanometer	['nænəmiːtə]	*n.*	纳米
phasor	[feɪzə(r)]	*n.*	相量，相量图
repetitive	[rɪ'petɪtɪv]	*adj.*	重复，迭代
sinusoid	['saɪnəsɔɪd]	*n.*	正弦曲线，正弦

amplitude modulation	(AM)调幅
analogous to	类似于……
frequency modulation	(FM)调频
lag network	滞后网络
lead network	超前网络
refer to…as	提到……，作为……，把……称为
respond to	对……响应
transient analysis	暂态分析

Notes to the Text

1. Gustav Kirchhoff 人名，德国物理学家基尔霍夫（1824—1887年），他给出了电路分析的两个基本定律
2. vice versa 反之也一样
3. time-invariant circuit 时不变电路（稳态电路）

6.2 Reading Materials

6.2.1 Circuit Breaker

The circuit breaker (Fig 6.5(a)) is an absolutely essential device in the modern world, and one of the most important safety mechanisms in your home. Whenever electrical wiring in a building has too much current flowing through it, these simple machines cut the power until somebody can fix the problem.

The basic circuit breaker consists of a simple switch, connected to either a bimetallic strip or an electromagnet. Fig 6.5(b) shows a typical electromagnet design.

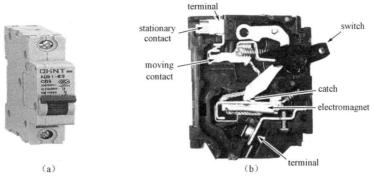

Fig 6.5　circuit breaker

The hot wire in the circuit connects to the two ends of the switch. When the switch is flipped to the on position, electricity can flow from the bottom terminal, through the electromagnet, up to the moving contact, across to the stationary contact and out to the upper terminal.

The electricity magnetizes the electromagnet. Increasing current boosts the electromagnet's magnetic force, and decreasing current lowers the magnetism. When the current jumps to unsafe levels, the electromagnet is strong enough to pull down a metal lever connected to the switch linkage. The entire linkage shifts, tilting the moving contact away from the stationary contact to break the circuit. The electricity shuts off.

More advanced circuit breakers use electronic components (semiconductor devices) to monitor current levels rather than simple electrical devices. These elements are a lot more precise, and they shut down the circuit more quickly, but they are also a lot more expensive.

6.2.2　Information on Amplitude Modulation (AM)

Less power is required to transmit high frequencies than low frequencies, therefore, it is more efficient to transmit high frequencies containing information over great distances. In amplitude modulation (AM), a constant amplitude radio frequency (RF) carrier wave of, for example, 800 kHz is produced. The information to be transmitted, perhaps an audio frequency (AF) of 50 Hz, is sent to a modulating circuit, where it varies the amplitude of the RF carrier wave. An AM signal as seen on an oscilloscope will show modulation peaks and modulation valleys. The amount of RF carrier wave variation depends on the amplitude of the audio signal and is called percentage of modulation (Fig 6.6). The loudness of the information at the receiver is a result of the percentage of modulation). Usually, AM transmission is kept below the 100% level of modulation.

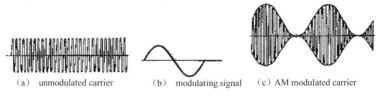

Fig 6.6　amplitude modulation

6.2.3 Thévenin's Theorem

Suppose that a load resistor R_L is connected to an arbitrary (in the sense that it contains only linear elements) circuit as shown in Fig 6.7(a). What value of the load R_L will absorb the maximum amount of power? Knowing the particular circuit, we can use nodal or mesh analysis to obtain an expression for the power absorbed by R_L, then take the derivative of this expression to determine what value of R_L results in maximum power. The effort required for such an approach can be quite great. Fortunately, though, a remarkable and important circuit theory concept states that as far as R_L is concerned, the arbitrary circuit shown in Fig 6.7(a) behaves as though it is a single independent voltage source in series with a single resistance.

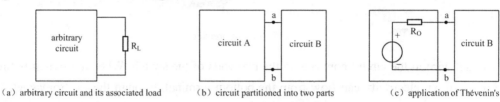

(a) arbitrary circuit and its associated load　　(b) circuit partitioned into two parts　　(c) application of Thévenin's

Fig 6.7　Thévenin's Theorem

Once we determine the values of this source and this resistance, we simply apply the results on maximum power transfer.

Suppose we are given an arbitrary circuit containing any or all of the following elements resistors, voltage sources, and current sources. (The sources can be dependent as well as independent.) Let us identify a pair of nodes, say node a and node b, so that the circuit can be partitioned into two pars as shown in Fig 6.7 (b), Furthermore suppose that circuit A contains no dependent source that is dependent on a variable in circuit B, and vice versa. Then we can replace circuit A by an appropriate independent voltage source, call it u_{OC}, in series with an appropriate resistance, call it R_O, and the effect on circuit B is the same as that produced by circuit A. This voltage source and resistance series combination is called the Thévenin equivalent of the circuit A. In other words, circuit A in Fig 6.7 (b) and the circuit in the left box in Fig 6.7(c) have the same effect on the circuit B. This result is known as Thévenin's theorem, and is one of the more useful and significant concepts in circuit theory.

To obtain the voltage u_{OC} - called the open-circuit voltage-remove circuit B from circuit A and determine the voltage between nodes a and b. This voltage, as shown in Fig 6.8 (a) is u_{OC}.

To obtain the resistance R_O - called the Thévening-equivalent resistance or the output resistance of circuit A - again remove circuit B from circuit A. Next, set all independent sources in circuit A to zero, Leave the dependent sources as is! (A zero voltage source is equivalent a short circuit, and a zero current source is equivalent to an open circuit) Now determine the resistance between nodes a and b - this is R_O as shown in Fig 6.8 (b).

If circuit A contains no dependent sources, when all independent sources are set to zero, the result may be simply a series-parallel resistive network in this case, however, R_O can be found by applying an independent source between nodes a and b and then by taking the ratio of voltage to current. This procedure is depicted in Fig 6.8(c), for the most part it doesn't matter whether u_O is applied and i_O is calculated or vice versa.

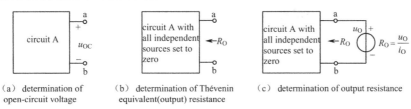

(a) determination of open-circuit voltage

(b) determination of Thévenin equivalent(output) resistance

(c) determination of output resistance

Fig 6.8　to obtain the resistance R_O

6.2.4　Apple iPod Classic 120 GB Silver

Now you can take it with you. All of it. Available in a 120 GB model that holds up to 30 000 songs, 150 hours of video, 25 000 photos, or any combination, the new iPod classic (Fig 6.9) fills your pocket with sight and sound. Available in quintessential silver or striking new black, iPod classic catches your eye with its sleek, all-metal enclosure composed of anodized aluminum and polished stainless steel.

Fig 6.9　Apple iPod

The new Genius Playlist feature creates an on-the-fly playlist of tracks in your library that go great with the song you're listening to. And Cover Flow lets you flip through your music by album artwork, or navigate your songs and videos by playlist, artist, album, genre, and more. You also can search for specific titles and artists. Want to mix things up? Click Shuffle Songs for a different experience every time.

Discovering new music, movies, TV shows, games, audiobooks, and podcasts is easy on the iTunes Store. To get everything into your pocket, just connect iPod classic to your Mac or PC, and iTunes transfers your music and more in one seamless sync.

6.3　Knowledge about Translation
（翻译知识 6——It 的用法）

1. it 作代词

① it 作无人称代词。

it 作无人称代词时可以表示自然现象、天气、时间、距离等，it 是形式上的主语，没有词汇意义，翻译时可省略。

② it 作人称代词。

it 作人称代词时，用来代替上文中提到的事或物，翻译时可译为它或译成它所代替的事或物。

The Zener diode maintains the voltage across its terminals by varying the current that flows through **it**. 稳压管通过改变流过它的电流来维持稳压管两端的电压（稳定）。

这里的 it 是指稳压管。

2. it 作形式主语

it 作形式主语时，可代替不定式短语或主语从句，这时 it 称为形式主语。

① it 代替不定式的句型有：

it is (was) ＋ 形容词＋不定式

it＋谓语动词＋不定式

It requires power to drive machines. 开动机器需要能源。

It 代替 to drive machines。

② it 代替主语从句的句型有：

it is ＋形容词＋主语从句

It is certain that	……是确定的。
It is clear that	……是很清楚的。
It is doubtful that	……是值得怀疑的。
It is desirable that	……是理想的。
It is essential that	……是必要的。
It is possible that	……是可能的。

it ＋谓语动词的被动态＋主语从句

It is well-known that	……是众所周知的。众所周知，……
It is said that	据说……
It is believed that	大家相信……
It is reported that	据报道……
It is generally recognized that	大家公认为……
It is supposed that	假设……
It is assumed that	假定……

It has been found that a force is needed to change the motion of a body. 要改变一个物体的运动（状态）需要加外力，这一点已被确定。

It is ＋名词＋主语从句

It is the case that	情况是……
It is a pity that	遗憾的是……
It is no use that	……是没用的
It is common knowledge that	常识是……

It ＋不及物动词＋主语从句

It now appears that	现在看来……
It seems that	好像是……
It turns out that	显然，……
It happens that	正巧……

3. it 作形式宾语

当动词不定式短语或从句在句中作宾语，而这种宾语又带有补足语时，通常要把 it 放在宾语补足语的前面，使语句简洁明了。

it 代替动词不定式短语

When we want to measure very small currents we find it convenient to use milliamperes and microamperes. 当要测量很小的电流时，我们觉得用毫安和微安是较为方便的。

it 代替宾语从句

The effects we have just discussed make it apparent that here is a means of converting mechanical energy into electrical energy. 从我们刚才讨论的结果来看，显然这是一种把机械能转换成电能的方法。

4. it 用于强调句型

强调句型是简单句，可以用来强调句中的主语、宾语和状语，但不能强调谓语和定语。强调语句的句型为：It is (was) ＋被强调的成分＋that (which, who)。在这种句型中，It 和 that 都没有意义，翻译时可在强调成分前加上"正是"、"就是"等。

It is in the form of alloys that metals are often used in industry. 在工业中就是经常以合金的形式使用金属。

6.4 Exercises

1. Put the Phrases into English

① 基尔霍夫电压定律
② 电压源
③ 电荷守恒定律
④ 在每一瞬时
⑤ 元件两端的电压
⑥ 无线电传输
⑦ 频率调制或调频
⑧ 频域
⑨ 线性电阻
⑩ 调幅波形

2. Put the Phrases into Chinese

① current source
② under this circumstance
③ present the second of Kirchhoff 's laws
④ introduce the concept of a "loop"
⑤ An alternative statement of KVL
⑥ voltages algebraically sum
⑦ sinusoidal steady-state response
⑧ ordinary household voltage
⑨ time-invariant circuit
⑩ percentage of modulation

3. Sentence Translation

① At any node of a circuit, at every instant of time, the sum of the currents into the node is equal to the sum of the sum of the currents out of node.

② Suppose that $u(t)$=1 V. Then the electric potential at the top of the material is 1 V above the potential at the bottom.

③ If the resulting current $i(t)$ is always directly proportional to the voltage for any function $u(t)$, the material is called a linear resistor.

④ If the input of a linear, time-invariant circuit is a sinusoid, then the response is sinusoid of the same frequency.

⑤ There are several distinct approaches that we can take. In one we write a set of simultaneous equations in which the variables are voltage, this is known as nodal analysis.

⑥ Less power is required to transmit high frequencies than low frequencies; therefore, it is more efficient to transmit high frequencies containing information over great distances.

⑦ An AM signal as seen on an oscilloscope will show modulation peaks and modulation valleys.

⑧ Furthermore suppose that circuit A contains no dependent source that is dependent on a variable in circuit B, and vice versa.

⑨ To obtain the voltage u_{OC} - called the open-circuit voltage-remove circuit B from circuit A, and determine the voltage between nodes a and b. This voltage is u_{OC}.

⑩ Suppose we are given an arbitrary circuit containing any or all of the following elements resistors, voltage sources, current sources.

4. Translation (Pay More Attention to What does *It* Mean)

① Resistors are used to limit current flowing to a device, thereby preventing **it** from burning out.

② An inductor is an electrical device, which can temporarily store electromagnetic energy in the field about **it** as long as current is flowing through **it**.

③ **It** is more efficient to transmit high frequencies containing information over great distances.

④ An inductor tends to oppose a change in electrical current, **it** has no resistance to dc current but has an ac resistance to ac current.

⑤ Let's insert a transistor in place of the switch to show how **it** can control the flow of electrons through the lamp.

6.5　课文参考译文

6.5.1　电路

当你揿下开关，打开灯、电视、真空吸尘器或计算机时，你有没有好奇地想到底里面是怎么回事？其实这时你正闭合（接通）一个电路，使电流（电子流）能够从导线中流过。

图 6.1 是一个手电筒的简单电路，其一端是电池，另一端是手电筒灯泡。当开关断开时，电路没有闭合，电路中没有电流，当开关合上时，构成了闭合电路，电流流过灯泡使它发光。

电路可能是庞大的电力系统，通过数千英里输送几百万瓦特的电能，也可能是含有数百万晶体管的微型芯片。就是有这些特别小的电路才有可能制造出台式计算机。最前沿的可能是纳米电子电路，其器件尺寸在纳米级（十亿分之一米）。

6.5.2　欧姆定律

如果把某种材料连接到一个理想电压源 $u(t)$ 的两端（如图 6.2 所示），假设 $u(t)$=1 V，则材料顶部的电势（电位）比其底部的电势高出 1 V，因为电子是携带负电荷的，在材料中的电子将从材料的底部流向顶部，即有电流 $i(t)$ 从材料的顶部通过材料流向底部。因此，当电源极性给定时，当 $u(t)$ 为正时，电流正方向如图 6.2 所示，如果 $u(t)$=2 V，同理材料顶部的电势（电位）比其底部的电势高出 2 V，所以电流 $i(t)$ 仍为正的（当材料为"线性"材料时，电流是原来的两倍)。对任意给定的电压函数 $u(t)$,如果产生的电流总是正比于给定的电压函数 $u(t)$,则该材料被称为线性电阻。

对一个线性电阻有:

$$R = \frac{u(t)}{i(t)} \quad \text{和} \quad i(t) = \frac{u(t)}{R}$$

电阻的单位(伏特每安培)被称做欧姆,并用大写希腊字母"Ω"表示,上面的两个方程都被称为欧姆定律。

6.5.3 基尔霍夫定律

德国物理学家基尔霍夫 Gustav Kirchhoff(1824—1887)得出了基尔霍夫定律,我们可以用它来分析任何电路元件(电压源、电流源和电阻)以及其他电子器件构成的相互连接。我们把这种相互连接称为一个电路或一个网络。

现在我们给出基尔霍夫两个定律中的第一个——基尔霍夫电流定律(KCL),它是根据电荷守恒的定律给出的:

对于一个给定的电路,两个或更多的元件的连接点称为节点(如图6.3所示)。对于电路的任何一个节点,在每一瞬时流入节点的电流之和总是等于流出节点的电流之和。

$$\sum i_{in} = \sum i_{out}$$

如果考虑电流的流向,取流入节点为正,流出节点为负,则基尔霍夫电流定律也可以等价表达成:

$$\sum i = 0$$

对于电路中的任一节点,电流的代数和为0。

现在给出基尔霍夫第二定律——电压定律。我们先引入"闭合回路"的概念。从电路的任一节点开始,通过电路中每一个元件再回到电路的出发节点且其他每个节点只能通过一次,形成一个闭合回路(如图6.4所示)。基尔霍夫电压定律(KVL)是:

$$\sum u_+ = \sum u_-$$

沿着电路的任一闭合回路,每一瞬时的(某一极性)上升电压之和等于(其相反极性)下降电压之和。

假定定义元件两端的电压极性从正到负为正电压,从负到正为负电压(反之亦然,即定义电压极性从负到正为正电压,从正到负为负电压也可以),基尔霍夫电压定律也可以表述成下列形式:

$$\sum u = 0$$

绕电路中的任一闭合回路,电压的代数和等于0。

6.5.4 电路分析方法

求出一个电路的变量(无论是电压还是电流)的过程称为电路分析。虽然一些简单的电路就已经十分有用了,对这些简单的单级放大电路,只要直接运用欧姆定律、KCL 和 KVL 就可以求解。但是我们必须学会分析更加复杂的电路,在复杂电路中显然不可能直接列出和求解线性代数方程组。

我们可采用几种不同的分析方法去分析复杂电路,其中一种是:列出一组瞬时方程,方程组中的变量全是电压,这种方法称为节点电压法。另一种方法是:列出一组变量全是电流的瞬时方程，这种方法称为网孔电流法。节点电压法可以适用于任何电路，但网孔电流法却不是，网孔电流法不适用于"非平面网络"。但还有一种与网孔电流法相似的，也以电流为变量的分析方法——回路电流法却可以适用于任何电路。

6.5.5 正弦电路

在讨论电路突然接通电源或电路的输入信号是突变或不规则的时候，可用阶跃函数和脉冲函数输入信号来分析电路的响应，这就是暂态分析过程。但如果想了解对一个有规律的或周期性的输入信号，电路是如何响应的，则到目前为止最有用的是正弦函数输入信号，这种分析称为稳态分析。

正弦函数是特别重要和常见的函数，首先日常用电的电压是（时间的）正弦（函数）的，其次无线电传输所用的信号不是调幅信号就是调频信号，调幅（AM）信号即一个其电压的幅度是按照一些信息信号的规律变化或称调制的正弦信号，调频（FM）信号即一个其频率被调制的正弦电压信号。

对于正弦电，我们有如下结论：

① 如果一个线性、时不变电路的输入是一个正弦信号，则它的响应是一个同频率的正弦信号。

② 通过用实数或复数求解我们可以求出一个正弦稳态响应的幅值和相位角。

③ 如果一个正弦电路的输出响应比输入信号先达到峰值，则称此电路是超前网络，否则就称为滞后网络。

④ 用相量和阻抗的概念，在同一频域的正弦电路可以采用类似于分析电阻电路的方法用基尔霍夫的电流、电压定律，节点电压分析法，网孔电流分析法和回路分析法的相量（复数）形式进行分析。

6.6 阅读材料参考译文

6.6.1 电路断路器

电路断路器（如图 6.5(a)所示）是当代最基本的电器，也是家中最重要的安全装置之一。当楼房中的电线中流过太大的电流时，这些简单的电器可以切断电源，当电路恢复正常后可以再次合上。

基本的电路断路器由一个简单的开关连接一个双金属片或一个电磁铁条。图 6.5(b)是一个典型的电磁铁型的断路器。

电路的相线（火线）与开关的两端相连接。当开关在接通的位置时，电流可以从底部端子绕过电磁铁，接在动触点，动触点与静触点相接触，最后到上部的端子（线路接通）。

电流使电磁铁产生电磁场，电流增大时电磁场的电磁力增大，电流减小时电磁力减小。当电流过大（达到不安全的水平，通常是超过所设定的最大允许电流时），电磁力大到可以把连接在开关联动机构上的一个金属杠杆向下拉开，联动机构脱构，使动触点离开静触点，断开电路，切断电流。

更先进的电路断路器不是用简单的电气部件，而是用电子器件（半导体器件）来监控电流，这些元件更加精确，切断电路时动作更快，但价格也贵很多。

6.6.2 关于调幅

传送高频信号要比传送低频信号（在传送过程的中）消耗的功率小，所以远距离传送包含信息的高频信号效率比较高。在幅度调制（调幅）时，先产生一个稳幅的无线电传送频率（射频）如 800 kHz 的载波信号，然后把要传送的信号，也许是一个 50 Hz 的音频信号送到一个调制电路中，通过调制电路使载波信号的幅度按音频信号的规律变化。在示波器中可看到一个调幅信号有调制峰值和调制谷底。射频载波（幅度）变化的大小取决于音频信号的幅度，称为调制百分比（如图 6.6 所示）。在接收器中信息（声音）的大小就取决于调制百分比。一般调幅波传输时其调制百分比小于 100%。

6.6.3 戴维南定理

设一个负载电阻 R_L 与一个任意电路（这个电路中只含有线性元件）相连接，如图 6.7(a)所示，那么电阻取何值时可以获得最多的电功率呢？对于一个具体的电路，我们可以用节点分析法或网孔分析法来求出 R_L 所获得（或消耗）功率的表达式。然后对这个表达式求导，求出电阻为多大值时消耗的功率最大。但用这种方法来求的话计算量比较大。然而一个著名的、重要的电路理论给出：如果只求与 R_L 相关的量时，图 6.7(a)中左边的任意电路可以用一个独立电压源串联一个电阻来表示。

只要我们求出了这个电压源和这个电阻的值，我们就可以很方便地分析电阻消耗的最大功率问题。

假设给定的任意电路中含有若干个（部分或全部）下列元件：电阻、电压源、电流源。（电源可以是受控源也可以是独立源。）我们在图上指定两个节点，称节点 a 和 b，则电路可以分成如图 6.7(b)所示的两部分。再假设电路 A 部分中没有受电路 B 部分中变量控制的受控电源，反过来也一样（即假设电路 B 部分中没有受电路 A 部分中变量控制的受控电源），则可用一个适当的独立电压源 u_{OC} 与一个适当的电阻 R_O 串联的电路来取代电路 A 部分，保证其对电路 B 的作用与电路 A（对电路 B 的作用）完全相同。这个电压源和串联电阻的组合称为电路 A 的戴维南等效电路。换言之，图 6.7 (b)中电路 A 和图 6.7 (c)中的左框中的电路对电路 B 有相同的作用。这个结果称为戴维南定理，是电路理论中比较有用和有意义的概念之一。

求开路端电压 u_{OC} 方法是从电路 A 旁边移走电路 B，如图 6.8(a)所示，求出节点 a 和 b 之间的电压就是 u_{OC}。

求戴维南等效电阻或称输出内电阻 R_O 的方法也是从电路 A 旁边移走电路 B，并将电路 A 中所有的独立电源除去，只留下受控电源（除源即把电路中的电压源等效成短路、电流源等效成开路），求出节点 a 和节点 b 之间的电阻就是 R_O，如图 6.8(b)所示。

如果电路 A 中不含受控电源，当除去所有的独立电源之后，得到的可能是一个简单的串并联电阻网络（可以直接求出 R_O）。另外，（尤其在电路 A 含受控电源时）R_O 也可以通过在节点 a 和节点 b 之间加一个独立电源，求出（a,b 两端）电压与（流入）电流之比来求得。这个过程如图 6.8(c)所示，在大部分情况中，加电压 u_O 求出电流 i_O 或者加电流 i_O 求出电压 u_O 都是可以的。

6.6.4 苹果（公司）iPod classic 120 GB 银色

你可以随身携带所有这一切：多达 120 GB 的存储容量，可以存储多达 30 000 首歌曲、150 小时的影片、25 000 张照片，或是上述这些数字媒体文件的任意组合，有了新一代的 iPod classic（如图 6.9 所示）就可把这些画面和歌曲放入口袋中。iPod classic 有精致银和闪耀黑两种颜色可选，其时髦的镀铝和抛光不锈钢的全金属外壳绝对吸引人的眼球。

全新的 Genius（直译是"天才"，此处可不译）功能能够在你的音乐库中找到与你正在听的歌曲口味类似的歌曲，并为你创建一个 Genius 播放列表。通过 Cover Flow 功能，你可以浏览专辑封面，或是通过播放列表、表演者姓名、专辑、歌曲类型等条件来选取歌曲和影片。此外，还可以利用搜索方式来找出特定的曲目或表演者。想自己混搭（组合）喜爱的歌曲吗？只要点选随机播放功能，每次都会有不同的惊喜体验。

可以很方便地在 iTunes 商店中找到新的音乐、影片、电视剧、游戏、有声读物和播客（一种可订阅下载音频文件的互联网服务，多为个人自发制作），想把所有这一切放入你的口袋吗？只要把 iPod classic 连接到 Mac（苹果电脑）或是 PC 上，iTunes 功能就会同步下载这些音乐或其他资料。

Unit 7　Integrated Circuit

Pre-reading

Read the following passage, paying attention to the question.
1) What is an Integrated Circuit?
2) What is ASIC?
3) What is chip holder?
4) What is a typical IC design process composed of?

7.1　Text

7.1.1　Information on Integrated Circuits

An integrated circuit (IC) is a small electronic device made out of a semiconductor material. The first integrated circuit was developed in the 1950s by Jack Kilby of Texas Instruments and Robert Noyce of Fairchild Semiconductor.

Integrated circuits (see in Fig 7.1) are used for a variety of devices, including microprocessors, audio and video equipment, and automobiles. Integrated circuits are often classified by the number of transistors and other electronic components they contain:

- SSI (small-scale integration): Up to 100 electronic components per chip.
- MSI (medium-scale integration): From 100 to 3 000 electronic components per chip.
- LSI (large-scale integration): From 3 000 to 100 000 electronic components per chip.
- VLSI (very large-scale integration): From 100 000 to 1 000 000 electronic components per chip.
- ULSI (ultra large-scale integration): More than 1 million electronic components per chip.

Fig 7.1　integrated circuits

Having an understanding of integrated circuits is becoming almost essential for an engineer doing system design. This is true because practically every modern system contains integrated circuits as critical subcomponents. As the capability to integrate a greater number of transistors in a single integrated circuit (IC) grows, it is becoming more common that an application-specific IC (ASIC) is required, at least for high volume applications.

Moore's law, which predicted that the number of devices integrated on a chip would be doubled every two years, was accurate for a number of years. Only recently has the level of integration begun to slow down somewhat due to the physical limits of integration technology. Advances in silicon technology have allowed IC designers to integrate more than a few million transistors on a chip; even a whole system of moderate complexity can now be implemented on a single chip.

To keep pace with the increasing complexity in very large-scale integrated (VLSI) circuits, the productivity of chip designers would have to increase at the same rate as the level of integration. Without such an increase in productivity, the design of complex systems might not be achievable within a reasonable time frame.

The rapidly increasing complexity of VLSI circuits has made design automation an absolute necessity, since the required increase in productivity can only be accomplished with the use of sophisticated design tools. Such tools also enable designers to perform trade-off analyses of different logic implementations and to make well-informed design decisions.

7.1.2 Chip and Chip Holders

Integrated circuits are usually called ICs or chips. They are complex circuits, which have been etched onto tiny chips of semiconductor (silicon). The chip is packaged in a plastic holder with pins spaced on a 0.1" (2.54 mm) grid, which will fit the holes on stripboard and breadboard. Very fine wires inside the package link the chip to the pins.

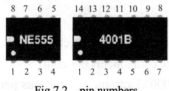

Fig 7.2 pin numbers

1. Pin Numbers

The pins are numbered anti-clockwise around the IC (chip) starting near the notch or dot. The Fig 7.2 shows the numbering for 8-pin and 14-pin ICs, but the principle is the same for all sizes.

2. Chip Holders (DIL Sockets)

ICs (chips) are easily damaged by heat when soldering and their short pins cannot be protected with a heat sink. Instead we use a chip holder (see in Fig 7.3), strictly called a DIL socket (DIL = Dual In-Line), which can be safely soldered onto the circuit board. The chip is pushed into the holder when all soldering is complete.

Chip holders are only needed when soldering so they are not used on breadboards.

Commercially produced circuit boards often have chips soldered directly to the board without a chip holder; usually this is done by a machine which is able to work very quickly. Please don't attempt to do this yourself because you are likely to destroy the chip and it will be difficult to remove without damage by de-soldering.

Fig 7.3 chip holders

3. Removing a Chip from Its Holder

If you need to remove a chip it can be gently prized out of the holder with a small flat-blade screwdriver. Carefully lever up each end by inserting the screwdriver blade between the chip and its holder and gently twisting the screwdriver. Take care to start lifting at both ends before you attempt to remove the chip, otherwise you will bend and possibly break the pins.

7.1.3 Bipolar Integrated Circuits & MOS Integrated Circuits

Historically, bipolar integrated circuits used to be much more popular than MOS integrated circuits, particularly for small-scale logic circuits. There were two major reasons for this: first, bipolar-junction transistors (BJT) originally could be manufactured more reliably than MOS transistors, and second, they were faster, As the reliability of MOS transistors improved, and as integrated circuits became more complex, which made the lower power and smaller size of MOS logic more important, the popularity of BJT logic decreased, however, BJT technology is still popular for the highest frequency logic circuits.

7.1.4 the Process of IC Design

A typical IC design process is composed of the four following categories:

① System (behavioral) design is the process of defining the circuit functionality and the input-output behavior. This level of specification may be expressed in terms of a flowchart or in terms of a high-level hardware description language (HDL).

② Logic design is the process of transforming a high-level description of a complex function into a net list of technology independent logic elements such as NAND gates, NOR gates, inverters and latches. This process helps to ensure that minimal logic is used to implement the function that was earlier defined in a high-level language.

③ Circuit design transforms the basic logic components into networks of transistors and interconnects.

④ Layout design creates geometrical shapes on various mask layers, which correspond to a silicon implementation of the circuit.

Although these steps are interrelated, each has its primary goals. At the system design level, the goal is to provide a complete and precise functional description. Logic-level design attempts to reduce the power consumption, and the objective of layout design is to realize circuit functions with a high packing density.

Technical Words and Phrases

chip	[tʃɪp]	*n.* 电路芯片，碎片，筹码　　*v.* 削成碎片，碎裂
delay	[dɪ'leɪ]	*vt.vi.* 推迟；延缓　　*n.* 延缓；迟延，耽误；阻塞；拖延
flowchart	[fləu'tʃɑːt]	流程图，程序框图
frame	[freɪm]	*n.* 框，框架；环境；背景
implement	['ɪmplɪmənt]	*n.* 工具；器具　　*vt.* 实现；履行
implementation	[ɪmplɪmen'teʃən]	*n.* 在系统开发过程中，用硬件、软件或其两者对一个系

		统设计的具体实现
integrate	['ɪntɪgreɪt]	vt. 集成，使成整体，使一体化，求……的积分 v. 结合
interrelate	[ɪntərɪ'leɪt]	vt.（使）互相联系
package	['pækɪdʒ]	n. 包，封装 vt. 打包，封装
pin	[pɪn]	n. 引脚，钉，销，栓，大头针，别针，腿 vt. 钉住，别住，阻止，扣牢，止住，牵制
screwdriver	[skru:draɪvə(r)]	n. 螺丝刀，改锥
solder	[səʊldə(r)]	n. 焊料，焊锡 vt. 焊接，焊补
sophisticated	[sə'fɪstɪkeɪtɪd]	adj. 高度发展的，精密复杂的，富有经验的；老练的
subcomponents	[sʌbkəm'pəʊnənt]	n. 子部件
well-informed	[welɪn'fɔ:md]	adj. 消息灵通的，熟悉的，博识的，见闻广博的

application-specific	专用的
be essential for/to…	对……来说是很重要的，很必要的
Chip holders	芯片插座
in terms of	以……的观点；就……而说
integrated circuit (IC)	集成电路
keep pace with…	与……保持一致的步伐
moderate complexity	中等规模（指（集成电路中）线路元件数及电路的复杂度为中等）
physical limit	物理条件限制（或译成硬件限制）
trade-off analyses	平衡分析，权衡利弊

Notes to the Text

Moore's law 摩尔定律

7.2　Reading Materials

7.2.1　Circuit Board

1. Breadboard (Temporary, no Soldering Required)

This is a way of making a temporary circuit, for testing purposes or to try out an idea. No soldering is required and all the components can be re-used afterwards. It is easy to change connections and replace components. Almost all projects started life on a breadboard (Fig 7.4(a)) to check that the circuit worked as intended.

2. Stripboard (Permanent, Soldered)

Stripboard (Fig 7.4(b)) has parallel strips of copper track on one side. The strips are 0.1" (2.54 mm) apart and there are holes every 0.1" (2.54 mm). Stripboard requires no special preparation other than cutting to size. It can be cut with a junior hacksaw, or simply snap it along the lines of holes by putting it over the edge of a bench or table and pushing hard.

(a) breadboard (b) stripboard

Fig 7.4　circuit board

3. Printed Circuit Board (Permanent, Soldered)

Printed Circuit Boards (PCB) (Fig 7.5) have copper tracks connecting the holes where the components are placed. They are designed specially for each circuit and make construction very easy. However, producing the PCB requires special equipment so this method is not recommended if you are a beginner unless the PCB is provided for you.

Fig 7.5　Printed Circuit Board (PCB)

7.2.2　Circuit Delay

Circuit delays in integrated circuits often have to be reduced to obtain faster response times. A typical digital integrated circuit consists of multiple stages of combinational logic blocks that lie between latches that are clocked by system clock signals. For such a circuit, delay reduction must ensure that valid signals are produced at each output latch of a combinational block, before any transition in the signal clocking the latch. In other words, the worst-case input-output delay of each combinational stage must be restricted to be below a certain specification.

Given the circuit topology, the delay of a combinational circuit can be controlled by varying the sizes of transistors in the circuit. Here, the size of a transistor is measured in terms of its channel width, since the channel lengths of MOS transistors in a digital circuit are generally uniform. In any case, what really matters is the ratio of channel width to channel length. In coarse terms, the circuit delay can usually be reduced by increasing the sizes of certain transistors in the circuit. Hence, making the circuit faster usually entails the penalty of increased circuit area. The area-delay trade-off involved here is, in essence, the problem of transistor size optimization.

7.2.3　3G Phones to Use Sony FeliCa IC Chip

Sony Corporation and NTT DoCoMo Inc. have agreed to form a Joint Venture Company to develop new services based on mobile phones equipped with Sony's contactless IC Card technology FeliCa(R).

Fig 7.6 shows how contactless communication between the reader/writer and the card is activated by electromagnetic waves radiated from the reader/writer.

Sony began developing FeliCa in 1988, and business use of the technology became widespread following the deployment of the technology in Hong Kong transport systems in 1997.

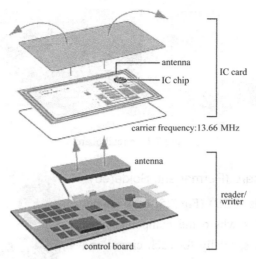

Fig 7.6　3G phones to use FeliCa IC chip

　　Other applications include the "Edy" electronic money service, the "eLIO" online credit service and various company/organizational ID security systems. At present, 38 million cards using FeliCa chips have been issued worldwide.

　　DoCoMo launched its mobile Internet service i-mode(R) in 1999, and has gained approximately 40 million subscribers to date. DoCoMo has constantly aimed to upgrade the service by adding various new technologies to enhance interface with other new devices, including infrared communications and two-dimensional bar codes. The enhancement of the i-mode handset interface is one primary example of DoMoCo's goal to promote mobile multimedia services by linking i-mode and real business.

　　The joint venture will develop the technology for a new IC chip, tentatively named "mobile FeliCa IC", which will integrate mobile phones with the FeliCa technology. The joint venture will work to create a platform whereby content providers can offer mobile services that feature both flexibility and security.

　　The mobile FeliCa IC and service platform will be developed in an open environment and provided to the widest possible range of mobile telecommunications operators and content providers.

　　Through this platform, customers will be able to use their mobile phones to enjoy services previously only available with IC cards, such as transport system payments, electronic money and personal ID security. In addition, use of the telecom infrastructure for mobile phones will also enable new capabilities. Potential applications of the technology could include electronic ticketing and online credit, which would provide convenience and added value for both providers and customers.

7.2.4　Electromagnetic Radiation and Lonosphere

　　The complete range of electromagnetic radiations is made up of gamma rays, X-rays, ultraviolet rays, ordinary visible light, infrared (heat) radiation, and radio waves. All these are given out by the Sun, but some of them do not reach us. As the atmosphere behaves like a filter, it lets through visible light and certain radio waves; most of the other radiation is absorbed. This is fortunate for us, because gamma rays, X-rays, and ultraviolet rays can harm living things.

A well-known characteristic of light is that it travels in straight lines. The other electromagnetic radiations behave in similar way. For this reason scientists thought Marconi Marhese Gugliemo (1874—1937), was wasting his time when he attempted to send radio waves from Cornwall to Newfoundland in 1902. It seemed obvious that the curvature of the Earth across the Atlantic would prevent this. However, the experiment was successful, and once again it had been shown that experiments should always be made to check predictions from theory.

A new theory was soon developed to explain the new facts; this was that a layer existed in the upper atmosphere, which could reflect radio waves. In this way they would be bounced round the world. This was called the Heaviside layer, after one of the scientists who predicted it; but as later work showed that there were several layers at different heights, they are now referred to by letters, and the general region in which they are present is called the ionosphere.

An atom of any substance consists of a positively-charged central body (the nucleus) surrounded by negatively-charged electrons, like planets around a sun. Each electron bears a single negative charge; and usually the nucleus carries a number of positive charges equal to the number of electrons. Therefore the whole atom is neither positive nor negative. If one or more electrons are removed, the remainder of the atom must be positively charged. It is called a positive ion. Sometimes it is possible to add extra electrons to the atom, and then we get negative ions.

Electric currents flowing in wires are streams of electrons running through the wire. Electrons also flow in a television tube, where they are made to hit the screen, causing a flash of light. Ions can also act as a current of electricity in liquids or gases, and thus the presence of electrons and ions in the upper atmosphere makes it electrically conducting. In fact, if there is a wind in the ionosphere, it will be an electric current.

7.2.5 Garmin Nuvi 260 3.5-Inch Portable GPS Navigator

Garmin nuvi 260 (Fig 7.7) combines the thin profile and attractive price point of other nuvi 200-series GPS with directions in real street names. As with all nuvis, you get Garmin reliability, the fast satellite lock of a high-sensitivity integrated receiver, a slim, pocket-sized navigator with a gorgeous display, detailed NAVTEQ maps that lets you search by name for more than 6 million points of interest like stores, restaurants or hospitals, and an easy, intuitive interface. Here are three features of nuvi 260.

1.Text-To-Speech

The text-to-speech feature of the nuvi 260 means that device automatically calls out street names (saying "turn right on Main Street" instead of "turn right in 200 feet."). This feature lets drivers keep their eyes on the road while navigating through busy traffic and tricky roadways.

Fig 7.7　GPS navigator

2. Smart, Powerful Design

The nuvi 260 W is built with a high-sensitivity GPS receiver for extreme accuracy, as well as an SD card slot for storing your media and additional navigation tools, and a USB interface for loading data. All this is wrapped up in a package that measures $4.8 \times 2.9 \times 0.8$ inches. The nuvi display is touchscreen-enabled, making it a cinch to control the device with your fingertips.

3. Navigate with Ease

The nuvi 260 comes ready to go right out of the box with preloaded City Navigator NT street maps, including a hefty POI database with hotels, restaurants, fuel, ATMs and more. Simply touch the color screen to enter a destination, and nuvi takes you there with 2D or 3D maps and turn-by-turn voice directions. In addition, the nuvi 260 accepts custom points of interest (POIs), such as school zones and safety cameras and lets you set proximity alerts to warn you of upcoming POIs.

（摘自 amazon 网站的广告）

7.3 Knowledge about Translation
（翻译知识 7——That 的用法）

that 的用法很多，在英语阅读和翻译中，正确区分 that 的作用，可以更好地理解一些复杂的句子。下面对 that 的主要用法做一介绍。

1. that 做指示代词

that 用做指示代词时，作定语相当于一个形容词修饰它后面的名词，作主语、表语和宾语时相当于一个名词，在科技文章中常用它来代替句中已出现过的某一名词，以免重复。

The proton has a single positive charge equal to **that** of an electron which is negative. 质子带有一个正电荷，其电量与电子所带的负电荷相等。

that 指代前面的 charge, which is negative 为定语，修饰 that 即 charge。

that 也可以用来指代整个句子。

Electrical energy can be changed easily and directly into all the other forms of energy. **That** is why electricity is so useful in everyday life. 电能可以方便而直接地转换成其他形式的能量，这就是电在日常生活中如此有用的原因。

that 指前面整个句子。

2. that 作关系代词引出定语从句

that 用做关系代词引出定语从句与 that 引出其他从句在语法结构上不同，that 在引出的定语从句中是一个语法成分，作主语或宾语。

An element is a simple substance **that** cannot be broken up into anything simpler. 元素是一种单质，它不可能再分成任何更简单的物质。

其中 that 在从句中为主语，指 element。

This process helps to ensure that minimal logic is used to implement the function **that** was earlier defined in a high-level language. 这个过程有助于保证用最小化的逻辑来实现前面用高级语言定义的功能。

其中 that 在从句中为主语，指 function。

3. that 作为连词引出名词性从句

这时 that 本身无词汇意义，也无复数形式，只在句中起语法上的连接作用。

It is becoming more common **that** an application-specific IC (ASIC) is required, at least for high volume applications. 专用集成电路的需要已更加普遍，至少对大批量的应用器件来说更需要专用的集成电路。

这里因主语很长，用 it 作形式主语，用 that 引出主语从句。

A well-known characteristic of light is **that** it travels in straight lines. 光的一个众所周知的特征是它以直线传播。

有一些名词常用 that 引出的同位语从句来加以说明，这样的名词有：fact（事实），idea（思想，概念），theory（理论），conclusion（结论），discover（发现）等。

From this we come to the conclusion **that** the resistance depends on the sort of material of which the electric conductor is made. 由此我们得出结论，电阻取决于导体是由哪种材料做成的。

4．in that 的用法

Liquids are different from solids **in that** liquids have no definite shape. 液体与固体的区别在于，液体没有固定的形状。

这里 that 作 in 的宾语，注意一般在介词后面是不能用 that 的，但 in that 是个特例，可以记成固定用法，译为：因为……，表示一些内在的原因。

5．连词引出状语从句

常用的结构形式有：so that…（导致，结果是）， so＋形容词（或副词）＋that（如此……以至于）, such that … 或 such＋名词＋that（这样……以至于）引出结果状语从句。

Some atoms are so constructed **that** they lose electrons easily. 有些原子的结构使它很容易失去电子。（如此构建以至于……）

Porcelain is such a good insulator **that** it is widely used. 瓷是一种很好的绝缘体，因而得到了广泛的应用。

Without such an increase in productivity **that** the design of complex systems might not be achievable within a reasonable time frame. 如果没有提高芯片设计的生产力（率），就不可能在一个合理的时间段内实现复杂系统的设计。

7.4 Exercises

1. Put the Phrases into English

① 数字集成电路
② 系统设计
③ 专用集成电路
④ 芯片设计能力（生产力）
⑤ 精致复杂的设计工具
⑥ 快速时间响应
⑦ 组合逻辑（电路）块
⑧ 有效信号
⑨ 输入输出延迟
⑩ 高频逻辑电路

2. Put the Phrases into Chinese

① very large scale integrated (VLSI) circuits

② in coarse terms
③ integrate a greater number of transistors in a single die or integrated circuit (IC)
④ make well-informed design decisions.
⑤ the ratio of channel width to channel length
⑥ output latch of a combinational block
⑦ defined in a high-level language
⑧ reliability of MOS transistor
⑨ reduce the power consumption
⑩ to realize circuit functions

3. Sentence Translation

① Moore's law, which predicted that the number of devices integrated on a chip would be doubled every two years, was accurate for a number of years.

② Advances in silicon technology have allowed IC designers to integrate more than a few million transistors on a chip; even a whole system of moderate complexity can now be implemented on a single chip.

③ Given the circuit topology, the delay of a combinational circuit can be controlled by varying the sizes of transistors in the circuit.

④ However, producing the PCB requires special equipment so this method is not recommended if you are a beginner unless the PCB is provided for you.

⑤ In any case, what really matters is the ratio of channel width to channel length.

⑥ Sony Corporation and NTT DoCoMo. Inc have agreed to form a Joint Venture Company to develop new services based on mobile phones equipped with Sony's contactless IC Card technology FeliCa(R).

⑦ Through this platform, customers will be able to use their mobile phones to enjoy services previously only available with IC cards.

⑧ This is fortunate for us, because gamma rays, X-rays, and ultraviolet rays can harm living things.

⑨ Electrons also flow in a television tube, where they are made to hit the screen, causing a flash of light.

⑩ Potential applications of the technology could include electronic ticketing and online credit, which would provide convenience and added value for both providers and customers.

4. Translation (Pay More Attention to that)

① Generally an electronic power supply is considered as a device **that** converts AC into DC.

② No matter what it is **that** spins the generator, commercial electrical generators of any size generate what is called 3-phase AC power.

③ Stealing a notebook computer from your competitor in the future, you may be surprised to find **that** the data on its hard drive has self-destructed in your hands.

④ The transformer's job is to reduce the 7 200 volts down to the 240 volts **that** makes up normal household electrical service.

⑤ For more than three decades, the power of microprocessors has doubled every 18 to 24 months, and most observers expect **that** to continue for another 10 years or so.

7.5 课文参考译文

7.5.1 关于集成电路

集成电路是用半导体材料制成的小电子器件，第一块集成电路是在20世纪50年代由德克萨斯仪器公司的Jack.Kilby和Fairchild半导体公司的Robert.Noyce开发的。

在各种设备包括微处理器、音频和视频设备以及汽车中都要用到集成电路（如图7.1所示）。集成电路通常根据其包含的晶体管和其他电路元件的数量来归类。

SSI（小规模集成电路）：每个芯片中有100个以下的电子元件。

MSI（中规模集成电路）：每个芯片中有100到3 000个的电子元件。

LSI（大规模集成电路）：每个芯片中有3 000到100 000个的电子元件。

VLSI（很大规模集成电路）：每个芯片中有100 000到1 000 000个的电子元件。

ULSI（超大规模集成电路）：每个芯片中有100万个以上的电子元件。

对一个进行系统设计的工程师来说了解集成电路是必不可少的。因为实际上现在每个系统中都含有作为关键部件的集成电路。随着在一个芯片上集成大量晶体管的能力（即集成电路的集成量）的提高，对专用集成电路的需求已更加普遍。至少对大批量的应用来说更需要专用的集成电路。

摩尔定律指出：在一个芯片上器件集成的数量将每两年翻一番，这些年来（集成电路的发展）是符合摩尔定律的。只是最近集成电路的发展水平开始变缓，多少是由于集成技术在物理上的限制（造成的）。硅芯片技术的发展使得集成电路设计者可以在一个芯片上集成几百万个以上的晶体管，甚至现在可以在单一芯片上集成一个中等复杂的系统。

为了与很大规模集成电路（VLSI）的复杂度的不断增加相适应，芯片设计的生产力（如设计者的人数，设计的手段等）也要不断提高，与集成水平保持相同的（发展）速度。如果在生产力方面没有这样一个提高，复杂系统的设计将不能达到在一个合理的时间结构中（同步发展）的要求。

VLSI电路复杂度的迅速增加，绝对有必要使得VISI设计成为自动化，因为只有利用精密复杂的设计工具才能实现生产力的不断提高。这些设计工具还可以使设计者对不同的逻辑设计的具体实现进行综合分析，从而做出全面的设计决定。

7.5.2 芯片和芯片插座

集成电路是把复杂电路刻蚀到微小的半导体硅芯片上，故常称做芯片，双列直插式芯片是用塑料封装且两侧带有间隔为0.1英寸（2.54 mm）的引脚，这些引脚可以插在焊接板上或面包板上，封装内部有很细的线把这些引脚与芯片相连。

1. 引脚的数目

引脚是用数字从一个缺口或一个圆孔开始逆时针编号的，图7.2显示一个8脚和一个14脚的芯片的引脚编号，对所有的尺寸芯片都是这样编号的。

2. 芯片插座

芯片焊接时受热易损坏，它们短短的引脚也不能用散热片保护，我们用芯片插座来保护（如图 7.3 所示），严格上应称为双列直插插座，可以先把插座焊接到电路板上，再把芯片插入插座。

只有当需要焊接时才用芯片插座，用面包板搭电路时不用插座。

实际产品的电路中通常不用插座而是把芯片直接焊在电路板上，因为实际产品是用机器焊接的，焊接速度很快（不会损坏芯片）。你不要这样做，因为可能会损坏芯片，而且在调试时要移去芯片，而焊住的芯片要取下且不损坏芯片是很难的。

3. 从插座上拔出芯片

如果你要取下一个芯片，你可以用一个小平口螺丝刀轻轻地把芯片从插座中拔出。仔细地在两端把平口螺丝刀插入芯片和插座之间，轻轻地翘起芯片，注意要分别在两端口起翘，否则会把引脚弄弯甚至折断。

7.5.3 双极型（晶体管）集成电路和 MOS 集成电路

历史上，双极型集成电路曾经比 MOS 集成电路更普及，尤其是对小规模集成逻辑电路而言。主要的原因有两个，一是最初双结晶体管（即三极管）要比 MOS 管更可靠，其次是三极管的速度比 MOS 管快。但随着 MOS 管可靠性的改进以及集成电路变得越来越复杂，使得 MOS 逻辑电路的体积小及（损耗）功率小这两个优点显得更重要了，双极型逻辑电路的使用也减少了，然而，双极型晶体管（集成电路）技术在高频逻辑电路中仍然很有用。

7.5.4 集成电路的设计过程

一个典型的集成电路设计过程是由 4 个部分组成。

① 系统（性能）设计是定义电路功能和输入输出特性的过程。这一级的设计说明可以用流程图或用一种高级硬件描述语言（HDL）来表示。

② 逻辑设计是把一个复杂功能的高级描述（定义）转换成由独立逻辑部件如与非门、非门、转换器、锁存器构成的功能列表，这个过程有助于保证用最小化的逻辑来实现前面高级语言定义的功能。

③ 电路设计是把基本逻辑部件转换成晶体管和内部连接构成的电路的过程。

④ 线路板设计是根据硅材料电路的需要在各层面上创建几何形状（即制作具体的晶体管元件等）。

虽然这些步骤相互之间是有联系的，但每一步都有它自己的基本目标。在系统设计阶段，目标是提供一个完整和精确的功能描述。逻辑阶段的设计目标是减小能量消耗，线路板设计目标则是以高集成度实现一个电路的功能。

7.6 阅读材料参考译文

7.6.1 电路板

1. 面包板（临时的，不需焊接）

面包板是在需要测试电路和验证一个（设计）想法时用来搭建一个临时电路的。用面包板搭电路不要焊接，所有的元器件都可以反复再用，可以很方便地改变电路连接及更换元器件。几乎所有的设计都是先用面包板（图 7.4(a)）来检查电路是否符合设计要求的。

2. 条形焊接板（永久的，需焊接）

条形焊接板（图 7.4(b)）有一面有平行的铜条，铜条的间距是 0.1 英寸（2.54 mm），铜条上面每隔 0.1 英寸（2.54 mm）有小孔。条形焊接板只要取适当大小就可以用了，可以用一般的钢锯锯开，也可以简单地把它放在长凳、桌子边上，沿着一排有孔的直线用力折断。

3. 印刷电路板（PCB）（永久的，需焊接）

印刷电路板（PCB）（如图 7.5 所示）用铜条（线）连接到各个器件的焊接孔上，每块电路板都是专门设计的，焊接比较容易。但制作 PCB 板需要专用的设备，所以不推荐新手（初学者）用 PCB 板，除非已有现成的 PCB 板。

7.6.2　电路延迟

集成电路中的电路延迟经常会影响电路快速响应时间，一个典型数字集成电路通常由多级组合逻辑块组合而成，各级逻辑块由锁存器锁存输入输出信号，而锁存器是用系统时钟信号作为脉冲触发信号的。对这样一个电路，必须保证延迟量不会影响在信号脉冲触发锁存器转换之前，每一个组合块的输出锁存器上已经产生一个有效的信号。换言之，要保证在最糟糕的情况下，每级组合块的输入输出延迟时间低于一个定值。

在已知电路拓扑（结构）条件下，组合电路的延迟可以通过改变在电路中晶体管的尺寸来控制，因为 MOS 晶体管的沟道长度通常是标准化（一定）的，这里晶体管的尺寸是由晶体管的沟道宽来确定的，所以通常（晶体管的尺寸）实际是指沟道宽度与沟道长度的比值。简言之，可以通过增加电路中晶体管的尺寸来减少电路的延迟（时间）。因此制造快速集成电路就需要增加电路的面积，这里所涉及的面积和电路延迟的矛盾实质上就是晶体管设计优化问题。

7.6.3　用索尼公司 FeliCa 芯片的 3G 手机

索尼公司和 DoCoMo 日本电话电报公司已达成协议建立一个合资公司开发一项基于装有索尼非接触 IC 卡的移动电话的新业务。

图 7.6 显示在发送/接收和卡之间的非接触通信是如何通过读/写设备发射的无线电磁波来激活（实现）的。

索尼公司在 1988 年开始研制非接触智能卡技术，自 1997 年中国香港的运输系统采纳这种技术后，这种技术得到广泛的商业应用。

其他的应用有：EDY 电子货币服务系统、eLIO 在线信用服务系统和各种公司/组织 ID 保密系统。目前，已有 3 800 万张带有 FeliCa 芯片的卡在世界各地流通。

DoCoMo 公司在 1999 年开办了它的网络移动电话系统 i-mode(R)，到目前已有近 4 000 万注册用户。DoCoMo 公司一直致力于升级其服务系统，通过采用各种新技术来增强它与其他新设备的接口能力，包括红外通信和二维条形码技术。网络移动电话手机接口技术的提高也是 DoCoMo 公司的主要目标之一，其目的是通过连接网络移动电话系统与实际商务，提高移动多媒体服务（功能）。

合资公司将开发新的 IC 芯片技术，暂称做"移动 FeliCa 卡"，把集成移动电话与 FeliCa 技术相结合，合资公司将创建一个平台，使信息提供商可以提供同时具有机动和保密特点的移动服务。

移动 FeliCa 卡和服务平台将在一个开放式环境中开发，在最大可能范围内提供给移动通信用户和信息提供商。

通过这个平台，用户可以用他们的移动电话享受原先只能由 IC 卡提供的服务，如公交卡、电子付费和个人身分证卡。再加上利用电信的移动电话的基本设施，还将实现新的功能。这种技术的潜在应用可能包括电子票和网上信用卡业务，这些将给服务商和用户提供方便。

注：**3G** 为 3rd Generation 的缩写，指第三代移动通信技术。相对第一代模拟制式手机(1G)和第二代 GSM、TDMA 等数字手机(2G)，第三代手机是指将无线通信与因特网等多媒体通信结合的新一代移动通信系统。它能够处理多种媒体形式，提供包括网页浏览、电话会议、电子商务等多种信息服务。

FeliCa 是由索尼公司开发的一种非接触智能卡技术。

"**EDY**"卡是一种嵌入到集成电路中以存储数据的电子货币卡。

7.6.4　电磁辐射和电离层

电射辐射波的完整范围是由伽马射线、X 射线、紫外线、普通可见光、红外（热）辐射和无线电波组成，所有这些电磁辐射是由太阳发出的，但其中有一些波并不能到达我们（地球）。因为大气层像一个滤波器，它只让可见光和一些无线电波通过，其他许多辐射都被大气层吸收了。这对我们来说是一件好事，因为伽马射线、X 射线和紫外线对生物有害。

光的一个众所周知的特征是光以直线传播，其他电磁波也有同样的特征。因此在 1902 年虽然有些科学家认为马可尼（1874—1937 年，意大利无线电报发明者）想从康沃尔发射无线电波到纽芬兰是在浪费时间，因为似乎很显然，跨越大西洋的地球弧形曲线使这种发射成为不可能。然而，实验却成功了，它再次证明了实验总是可以用来检验理论得出的预言。

为解释这个新的实验事实，一个新的理论很快被建立起来，这就是大气层上方存在一个层，它能反射无线电波，所以无线电波可以（弹）返回到地面，这个层称为电离层。在第一个科学家预言了电离层的存在后，接下来的研究工作表明在不同的高度有几个层，现在（分别）用字母来表示它们，这几个层所处的范围一般统称为电离层。

任何物质的原子是由一个带正电荷的中心体（原子核）和绕在它周围的带负电的电子组成，就像行星绕太阳一样，通常电子带有一个负电荷（电子），而原子核带有与电子数目相等的正电荷，所以整个原子是既非正也非负的（电中性的）。如果一个或多个电子脱离了原子核，则剩下的原子是带正电的，称为正离子。有时也有可能一个原子带上额外的电子，则我们称它为负离子。

电流在一个导线中流动就是电子在导线中移动形成的电子流。（注：但电流的流向与电子流的流向相反。）电子也可以在一个电视机的显像管中流动，在显像管中电子撞击显示屏，产生一个光（闪烁）点，在液体和气体中离子也可以与电子流一样流动，因此在大气层上部存在的电子和离子使大气层成为一个导电体，实际上如果电离层中有一阵风，这阵风就是一个电流。

7.6.5　Garmin Nuvi 260 3.5 英寸便携式导航仪

Garmin（美国一公司的名称）生产的 nuvi 260 导航仪（如图 7.7 所示）不但与其他 nuvi 200 系列导航仪一样纤薄，价格吸引人，而且会播报实际路名。同其他所有的 nuvi 系列导航仪一样，Garmin 公司的产品质量可靠，高灵敏集成接收器能快速锁定卫星，导航仪纤薄，口袋大小却有漂亮的显示屏，详细的导航地图可以让你通过搜索名字找到多达 600 万个地方，例如商店、旅馆或医院，人机界面直观。下面是 nuvi 260 的 3 个特点。

1．文本转声音

nuvi 260 的文本转声音特点是指它能自动报出街道名（例如"右转到主街道"而不是"右转 200 尺"）。这个特点使驾驶员在通过交通拥挤的道路时能看着路面而不是导航仪。

2．聪明、功能强大的设计

nuvi 260W 内置一个高灵敏 GPS 精密接收器，另有 SD 卡槽可以（插卡，用来）存储媒体和附加的导航工具，一个 USB 接口可以下载资料。所有这些都封装在 4.8×2.9×0.8 英寸的盒子内。Nuvi 的显示屏是触摸式的，可以很方便地用手指去控制导航仪。

3．导航容易

nuvi 260 预置了（美国）城市导航街道地图，包括大量的饭店、旅馆、加油站、ATM 机等有用的地点数据，只要触摸彩色屏输入目的地，nuvi 就用二维或三维地图显示并在转弯时用语音报方向，nuvi 260 能报出顾客比较关心的地方（兴趣点）如学校区、有监控摄像等，让你设置接近的警告，提醒你即将要到达的兴趣点。

Unit 8　Digital Logic Circuits

 Pre-reading

Read the following passage, paying attention to the question.
1) Why we say the binary system is the primary language of the computer?
2) What is a NAND gate?
3) What is a flip-flop?

8.1　Text

8.1.1　Number Systems

The number system with which we are most familiar is the base 10, or decimal system. Recent technological developments have created the need for other number systems. The electronic computer, for example, required the development of systems that were easily adapted to electronic processes. These number systems were the binary (base 2), octal (base 8), and hexadecimal (base 16). The binary system is the primary language of the computer and the octal and hexadecimal systems are usually used for communication with the computer and for storage of information within the computer. Since computers can only process binary numbers or numbers coded in other systems such as octal and hexadecimal, the decimal system must be converted to one of these other systems before it can be processed by the computer. When the computer finishes its operations on the information, the output is printed or displayed in number system other than decimal, and this too must be converted, this time back to the decimal system.

Digital electronics is a "logical" science. Logic, generally speaking, is the science of formal principles of reasoning. Digital logic is the science of reasoning with numbers; a special circuit called a gate can perform nearly all-digital functions. If the logic operation is too complex for one gate, it can almost always be implemented through the use of a combination of gates. These extended logic circuits are called combinational logic.

8.1.2　Logical Gates

1. AND Gate and NAND Gate

Two of the most basic logic functions are the AND and NAND functions. The difference between these functions is that they are complements. This means they are opposite in function.

The AND gate provides a function in digital logic which gives a high output when all of its inputs are high. Fig 8.1(a) shows the symbol used to represent an AND gate. In this case, there are two inputs, A and B, and one output C. Gates with as many as eight inputs are available.

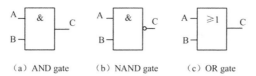

(a) AND gate　　(b) NAND gate　　(c) OR gate

Fig 8.1　symbol of gate

AND gate will provide an output high (1) only when both inputs (or all inputs if there are more than two) are high (1). This relationship is usually written in the following way:

$$C = AB$$

Above expressing is normally said in the following way: "if A equals 1 and B equals 1, then C will equal 1" or " A and B equal C".

The NAND gate is a more common gate than the AND gate (Figure 8.1(b)). It is the complement of the AND gate, meaning that it is just the opposite in logic state. NAND gates frequently are cheaper and easier to use than AND gates because electronically they are simpler. The N in front of the AND means "not" AND.

The logic is also written differently for an N gate when it is placed in a formula. An overscore of bar is used to indicate a NOT or inverted (complement) condition. For example:

$$C = \overline{AB}$$

It means "A and B equals NOT C." In the case of the NAND gate, the output sill be high(1) except when the inputs are all high. When all inputs are high, the output of the NAND gate will be low (0).

2. OR Gate

The OR Gate (Fig8.1(c)) provides a function that will give a high output (1) when any one of the inputs is at a high (1) logic level. This function is usually written in the following way:

$$C = A + B$$

The plus (+) sign is the symbol used to indicate an OR logic function. This expression is spoken in the following way: "if A OR B equals 1 then C equals 1."

3. Combining Logic Functions

Individual logic gates are building blocks. They can stand-alone if only a single logic function is needed, or they can be combined with other gates for more complex operations. There are times when it is advantageous to substitute one kind of gate for another. It is an indicator of the versatility and flexibility of digital integrated circuits that the designer can configure the gates on hand to meet the circuit needs.

8.1.3　the Flip-flops

1. the Clocked RS Latch

Microprocessors employ both latches and flip-flops. By adding a pair of NAND gates to the input circuits of the RS latch, we accomplish two goals: normal rather than inverted inputs, and a third input common to both gates which we can use to synchronize this circuit with others of its kind. The clocked RS NAND latch is shown in Fig 8.2(a).

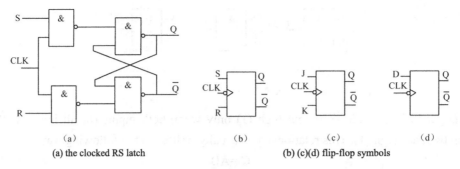

Fig 8.2　latch & flip-flop

(a) the clocked RS latch
(b) (c)(d) flip-flop symbols

The S and R inputs are normally at logic 0, and must be changed to logic 1 to change the state of the latch. However, with the third input, a new factor has been added. This input is typically designated C or CLK, because it is typically controlled by a clock circuit of some sort, which is used to synchronize several of these latch circuits with each other. The output can only change state while the CLK input is a logic 1. When CLK is a logic 0, the S and R inputs will have no effect.

The rule about not activating both the S and R input simultaneously holds true: if both are logic 1 when the clock is also logic 1, the latching action is bypassed and both outputs will go to logic 1. The difference in this case is that if the CLK input drops to logic 0 first, there is no question or doubt - a true race condition will exist, and you cannot tell which way the outputs will come to rest. The example circuit in Fig 8.2 (a) reflects this uncertainty.

2. the RS Flip-flop

The edge-triggered RS flip-flop actually consists of two identical RS latch circuits. The symbol of the RS flip-flop is shown in Fig 8.2 (b). The inverter connected between the two CLK inputs ensures that the two sections will be enabled during opposite half-cycles of the clock signal. This is the key to the operation of this circuit. Therefore, the Q and $\bar{Q}$ outputs can only change state when the CLK signal falls from a logic 1 to logic 0. This is known as the falling edge of the CLK signal; hence the designation edge-triggered flip-flop.

3. the JK Flip-flop

Because the behavior of the JK flip-flop is completely predictable under all conditions, this is the preferred type of flip-flop for most logic circuit designs.

The Q and $\bar{Q}$ outputs in a JK flip-flop will only change state on the falling edge of the CLK signal, and the J and K inputs will control the future output state.

If both the J and K inputs (Fig 8.2(c)) are held at logic 1 and the CLK signal continues to change, the Q and $\bar{Q}$ outputs will simply change state with each falling edge of the CLK signal. We can use this characteristic to advantage in a number of ways. A flip-flop built specifically to operate this way is typically designated as a T (for Toggle) flip-flop.

The JK flip-flop must be edge triggered in this manner. Any level-triggered JK latch circuit will oscillate rapidly if all three inputs are held at logic 1. This is not very useful. For the same reason, the T flip-flop must also be edge triggered. For both types, this is the only way to ensure that the flip-flop will change state only once on any given clock pulse.

4. the Edge-triggered D Flip-flop

One essential point about the D flip-flop(Fig 8.2(d)) is that when the clock input falls to logic 0 and the outputs can change state, the Q output always takes on the state of the D input at the moment of the clock edge. This was not true of the RS and JK flip-flops.

Technical Words and Phrases

asynchronous	[eɪˈsɪŋkrənəs]	adj. 异步的（时间不同的，不协调的）	n. 异步
binary	[baɪnərɪ]	adj. 二进位的，二元的	
code	[kəud]	n. 代码，代号，标记，法规 v. 编码	
configure	[kənˈfɪgə(r)]	vi. 配置，设定 vt. 使成形，使具一定形式	
decimal	[ˈdesɪm(ə)l]	adj. 小数的，以十为基础的 n. 小数，十进制	
extend	[ɪkˈstend]	v. 扩充，延伸，伸展，扩大	
flexibility	[flekˈsɪbɪlɪtɪ]	n. 灵活性，弹性，适应性	
flip-flop	[flɪpflɔp]	n. 触发器；触发电路，双稳态多谐振荡器	
gate	[geɪt]	n. 门（电路），栅极	
hexadecimal	[heksəˈdesɪm(ə)l]	adj. 十六进制的（0~9 和 A~F）	
individual	[ɪndɪˈvɪdjuəl]	adj. 单独的；特殊的；个别的	
latch	[lætʃ]	n. 寄存器，插销，撞锁，弹簧锁	
octal	[ˈɔktəl]	adj. 八进制的	
overscore	[əuvəskɔː(r)]	vt. 在（字、句等上或中间）画一条线 n. 字句等上或中间的线	
substitute	[ˈsʌbstɪtjuːt]	n. 替代人；替代物；代理人 vt.（常与 for 连用）替代	
synchronous	[ˈsɪŋkrənəs]	adj. 同时发生的，同步的	

combinational logic	组合逻辑
AND	"与"（计算机逻辑运算的一种，又称逻辑乘法）
AND gate	"与"门
NAND	"与非"，NAND gate "与非"门
OR	"或"（逻辑加）
N gate	"非"门，或写作 Not gate
edge-triggered flip-flop	边缘触发式触发器
Toggle	双态元件，触发器

8.2 Reading Materials

8.2.1 74 Series Logic ICs

There are several families of logic chips numbered from 74××00 onwards with letters (××) in the middle of the number to indicate the type of circuitry, e.g. 74LS00 and 74HC00. The original family (now obsolete) had no letters, e.g. 7400.

The 74LS (Low-power Schottky) family (like the original) uses TTL (Transistor-Transistor Logic) circuitry, which is fast but requires more power than later families. The 74 series is often still called the 'TTL series' even though the latest chips do not use TTL!

The 74HC family has High-speed CMOS circuitry, combining the speed of TTL with the very low power consumption of the 4000 series. They are CMOS chips with the same pin arrangements as the older 74LS family. Note that 74HC inputs cannot be reliably driven by 74LS outputs because the voltage ranges used for logic 0 are not quite compatible.

The 74HCT family is a special version of 74HC with 74LS TTL-compatible inputs so 74HCT can be safely mixed with 74LS in the same system. In fact 74HCT can be used as low-power direct replacements for the older 74LS ICs in most circuits. The minor disadvantage of 74HCT is a lower immunity to noise, but this is unlikely to be a problem in most situations.

8.2.2 Registers

A register is a group of flip-flops with each flip-flop capable of storing one bit of information. An n-bit register has a group of n flip-flops and is capable of storing any joinery information of n bits, in addition to the flip-flops, a register may have combinational gates that perform certain data-processing tasks. In its broadest definition, a register consists of a group of flip-flop and gates that effect their transition. The flip-flop hold the binary information and the gates control when and how new information is transferred into the register.

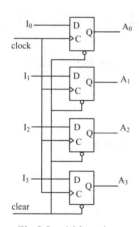

Fig 8.3 4-bit register

Various registers are available commercially. The simplest register is one that consists of only flip-flops, with no external gates. Fig 8.3 shows such a register constructed with four D flip-flops. The common clock input triggers all flip-flops on the rising edge of each pulse, and the binary data available at the four inputs are transferred into the 4-bit register. The four outputs can be sampled at any time to obtain the binary information stored in the register. The clear input goes to a special terminal in each flip-flop. When this input goes to 0, all flip-flops are reset asynchronously. The clear input is useful for clearing the register to all '0' prior to its clocked operation. The clear input must be maintained at logic 1 during normal clocked operation. Note that the clock signal enables the D input but that the clear input is independent of the clock.

The transfer of new information into a register is referred to as loading the register. If all the bits of the register are loaded simultaneously with a common clock pulse transition, we say that the loading is done in parallel. A clock transition applied to the clock input of the register of Fig 8.3 will load all four inputs I_0 through I_3 in parallel. In this configuration, the clock input must be inhibited from the circuit if the content of the register must be left unchanged.

8.2.3 Counter

All counters require a "square wave" clock signal to make them count.

1. Ripple and Synchronous Counters

There are two main types of counter: ripple and synchronous. In simple circuits their behavior appears almost identical, but their internal structure is very different.

A ripple counter contains a chain of flip-flops with the output of each one feeding the input of the next. A flip-flop output changes state every time the input changes from high to low (on the falling-edge). This simple arrangement works well, but there is a slight delay as the effect of the clock "ripples" through the chain of flip-flops.

The diagram (Fig 8.4) shows how to link ripple counters in a chain, notice how the highest output QD of each counter drives the clock input ($\overline{CK}$) of the next counter.

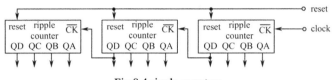

Fig 8.4 ripple counters

A synchronous counter has a more complex internal structure to ensure that all its outputs change precisely together on each clock pulse, avoiding the brief false counts, which occur with ripple counters.

The diagram (Fig 8.5) shows how to link synchronous counters, notice how all the clock (CK) inputs are linked. Carry out (CO) is used to feed the carry in (CI) of the next counter. Carry in (CI) of the first counter should be high.

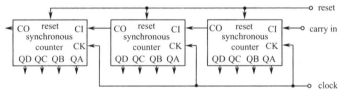

Fig 8.5 synchronous counter

2. Rising-edge and Falling-edge Clock Inputs

Counting occurs when the clock input changes state. Most synchronous counters count on the rising-edge, which is the low to high transition of the clock signal. Most ripple counters count on the falling-edge, which is the high to low transition of the clock signal (see in Fig 8.6).

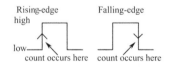

Fig 8.6 rising-edge & falling-edge

3. 74××390 Dual Decade (0-9) Ripple Counter

The 74××390 contains two separate decade (0 to 9) counters, one on each side of the chip (Fig 8.7). They are ripple counters so beware that glitches may occur in any logic gate systems connected to their outputs due to the slight delay before the later counter outputs respond to a clock pulse.

The count advances as the clock input becomes low (on the falling-edge), this is indicated by the bar over the clock label. This is the usual clock behavior of ripple counters and it means a counter output can directly drive the clock input of the next counter in a chain.

Each counter is in two sections: clockA-QA and clockB-QB-QC-QD. For normal use connect QA to clockB to link the two sections, and connect the external clock signal to clockA.

8.2.4 7-segment Display Drivers

The inputs A-D of a display driver are connected to the outputs QA-D from a decade counter (Fig 8.8). A network of logic gates inside the display driver makes its outputs a-g become high or low as appropriate to light the required segments a-g of a 7-segment display. A resistor is required in series with each segment to protect the LEDs, 330 Ω is a suitable value for many displays with a 4.5 V to 6 V supply. Beware that these resistors are sometimes omitted from circuit diagrams!

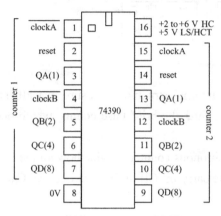

Fig 8.7   counter 74××390

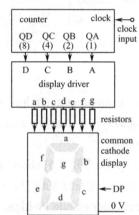

Fig 8.8   decade counters with display driver and 7-segment display

There are two types of 7-segment displays:

Common Anode (CA or SA) with all the LED anodes connected together and connect the common anode to $+U_S$. These need a display driver with outputs which become low to light each segment, for example, 74××47.

Common Cathode (CC or SC) with all the cathodes connected together and connect the common cathode to 0 V. These need a display driver with outputs which become high to light each segment, for example, 4511.

The common anode/cathode is often available on 2 pins. Displays also have a decimal point (DP) but this is not controlled by the display driver. The segments of larger displays have two LEDs in series.

8.3 Knowledge about Translation
（翻译知识 8——Which 的用法）

1. which 引导定语从句

which 用做关系代词时，引导出定语从句。这时它不仅起到把主句与从句连接起来的作用，而且用来代替句中某一名词。which 在从句中常用做主语、宾语或与介词连用，起状语作用。

① which 引导限定性定语从句。

限定性定语从句与它所说明的句中成分有密切联系，省去这种从句时，主句的意思就不能完整表达出来甚至主句根本不成立，这种从句不能用逗号与主句分开。

A conductor is a substance **which** is able to carry electrons easily. 导体是能很容易传输电子的物质。

The simplest atom is the hydrogen atom **which** contains one proton and one electron. 最简单的原子是氢原子，它只含有一个质子和一个电子。

② which 引导非限定性定语从句。

非限制性定语从句与它所说明的主句中的成分关系不密切，去掉非限定性定语从句一般不影响主句，翻译时可根据需要在从句前加上"因为"、"如果"、"由于"、"虽然"等相应的词。A current , **which** is constantly changing its direction through a circuit, is an alternating current. 通过电路时不断改变方向的电流称为交流电。which 引导的非限制性定语从句，进一步说明 current。

③ which 引导特种定语从句。

which 引导的特种定语从句通常用来代表主句的内容或部分内容，which 在从句中常常做主语，也可做定语或介词的宾语。做主语时谓语动词要用第三人称单数。

When the motor is turned on, a current flows in the wire **which** creates a sizable magnetic field around the wire coil. 当电动机转动时，导线中就有电流流动，这就产生了一个环绕着绕组线圈的相当大的磁场。（which 指 a current flows in the wire.）

2. "介词+which" 引导的定语从句

介词与 which 连用，在所引导定语从句中构成介词短语，主要用做状语。介词属于从句，分析时应当从介词开始断开。

The number system **with which** we are most familiar is the base 10, or decimal system. 我们最熟悉的数字系统是以 10 为基即十进制系统。

…amplitude modulation (AM), **in which** the amplitude of a sinusoid is changed or modulated according some information signal. ……调幅，在调幅时正弦信号的幅度是按一定信息信号变化或调制的。

这里 which 指前面的 amplitude modulation(AM)。

The strength of an electric current, for example, can be measured by the rate **at which** it flows through a wire. 电流的强度可以由电流流过导线的速率来测定。这里的 which 指的是 rate。

Magnetic field can be represented by magnetic lines of force, the direction **of which** at every point is that of the magnetic field. 磁场可以用磁力线表示，在每一点的磁力线的方向就是该点磁场的方向。

这里 which 指代 magnetic lines of force.

There are different forms of energy around us, any form **of which** can be changed into another form. 我们周围存在着不同形式的能量，其中任何形式的能量都能转变为另一种形式。

这里的 which 指代 energy.

3. which 用做连接代词

which 用做连接代词时，保留疑问含义，可引导主语从句，表语从句和宾语从句。通常译成"哪个"、"哪些"。

Which process will occur is determined entirely by the surroundings. 将出现哪种过程，完全取决于周围的环境。

4. which 用在强调句型中

It is (was) ＋ 强调成分＋which 这一强调句型中，只能强调句中的主语和宾语，但不能用来强调句中的状语。

8.4 Exercises

1. Put the Phrases into English

① 十进制数字系统
② 在计算机中信息的存储
③ 打印或显示
④ 逻辑运算
⑤ 组合逻辑门
⑥ 仅当输入全为高电平时输出为高电平
⑦ 上述表达式
⑧ 同步器件
⑨ 与其他门电路结合
⑩ 数字集成电路的通用性和多功能性

2. Put the Phrases into Chinese

① flip-flop
② asynchronous device
③ at arbitrary time
④ change state only on arrival of a clock pulse
⑤ the octal and hexadecimal systems
⑥ clear input
⑦ play video games
⑧ surf the web
⑨ with a common clock pulse transition
⑩ in its broadest definition

3. Sentence Translation

① The binary system is the primary language of the computer.

② Individual logic gates can stand-alone, if only a single logic function is needed, or they can be combined with other gates for more complex operations.

③ The 74HCT family is a special version of 74HC with 74LS TTL-compatible inputs so 74HCT can be safely mixed with 74LS in the same system.

④ A register is a group of flip-flops with each flip-flop capable of storing one bit of information.

⑤ The flip-flop hold the binary information and the gates control when and how new information is transferred into the register.

⑥ Logic, generally speaking, is the science of formal principles of reasoning.

⑦ When the computer finishes its operations on the information, the output is printed or displayed in number system other than decimal, and this too must be converted, this time back to the decimal system.

⑧ The clear input must be maintained at logic 1 during normal clocked operation.

⑨ A ripple counter contains a chain of flip-flops with the output of each one feeding the input of the next.

⑩ A synchronous counter has a more complex internal structure to ensure that all its outputs change precisely together on each clock pulse.

4. Translation (Pay More Attention to Which)

① In the reverse-biased condition, diode will not conduct until it reaches the zener voltage (U_z) at **which** it is designed.

② A modern robot is, in effect, a teleoperator in **which** the brain of the human being has replaced by a programmable microprocessors.

③ Moore's law, **which** predicted that the number of devices integrated on a chip would be doubled every two years, was accurate for a number of years.

④ Usually this is done by a machine, **which** is able to work very quickly.

⑤ Unlike most bipolar-junction transistor (BIJ) technologies, **which** make dominant use of only one type of transistor, MOS circuits normally use two complementary types of transistors-n-channel and p-channel.

8.5 课文参考译文

8.5.1 数字系统

我们最熟悉的数字系统是以 10 为基即十进制系统。但近年来科技的发展却需要产生其他的数字系统，如电子计算机需要开发可以适应电子处理的数字系统，这些数字系统是二进制（基为 2）、八进制（基为 8）和十六进制（基为 16）。二进制系统是计算机的基本语言，而八进制和十六进制系统通常在计算机通信和计算机存储信息时用。因为计算机只能处理二进制数据或其他系统如八进制和十六进制中的二进制编码，所以十进制系统必须先转换成这些系统才能用计算机进行处理。当计算机对所给的信息处理完毕后，输出的信息是以非十进制形式打印或显示出来，因此这些输出信息也必须再转换，这次是转换回十进制系统。

数字电子技术是一门逻辑科学，一般来说，逻辑学是按一般原则进行推理的科学。数字逻辑学是用数字推理的科学。几乎所有的数字（逻辑）功能都可以用一种特殊的称为门的电路来实现。如果逻辑运算太复杂，无法用一个门来实现，也可以通过几个门的组合来实现这个逻辑运算。这些扩展的逻辑电路被称为组合逻辑电路。

8.5.2 逻辑门

1. "与"门和"与非"门

两种最基本的逻辑函数是"与"和"与非"函数，这两个函数的差别在于它们是互补的，即在功能上是相反的。

"与"门的数字逻辑功能是当它所有的输入为高电平时它的输出为高电平。图 8.1(a)给出一个"与"门的符号，它有两个输入端 A，B 和一个输出端 C。门电路最多可以有 8 个输入端。

当"与"门的两个输入（或当它的输入不止两个时，所有的输入）为高电平（1）时它输出高电平（1）。这个关系通常可以写做：

$$C = AB$$

这个表达式可以说成：当 A 等于 1 且 B 等于 1 时有 C 等于 1 或 A 与 B 等于 C。

"与非门"比"与"门用得更普遍（图 8.1(b)），它和"与"门是互补的，即逻辑上它的输出和"与"门相反。因为"与非"门的电路实现比较简单，所以它比"与"门便宜，使用也方便。在 AND 前面的字母 N 意味着是非"与"。

逻辑（表达式）中对一个非门也有不同的表示法，常用字母上加一个上画线来表示非的关系。如：

$$C = \overline{AB}$$

它表示"A 与 B 等于非 C"。在与非门中，大部分情况下它输出都为高（1），除非当所有的输入都为高时，与非门的输出为低（0）。

2. "或"门

"或"门（图 8.1(c)）的逻辑功能是指当"或"门的任何一个输入为逻辑上高电平（1）时它输出为高（1），这种逻辑功能通常可写成：

$$C = A + B$$

加号"(+)"意味着"或"这种逻辑关系，可说成："如果 A 或 B 等于 1 则 C 等于 1"。

3. 组合逻辑门

单个逻辑门是就像积木（构造块），如果只需要一个单一逻辑关系时可以只用它们其中一个，但要实现更复杂的逻辑运算时它们也可以与其他门结合构成组合逻辑门。有时用一种门去代替其他的门更有利，设计者可以用手边的门电路来实现各种电路，这就是数字集成门电路的多功能性和灵活机动性。

8.5.3 触发器

1. 可控 RS 锁存器

微处理器要用到锁存器和触发器，通过在 RS 锁存电路的输入端加一对"与非"门，可以达到两个目的：一是不用反相输入，二是两个"与非"门有一个公共输入端。这个公共输入端可以使这个电路与其他同类电路同步，这就是如图 8.2(a)所示的可控 RS "与非"门锁存器。

R 和 S 的输入端通常接逻辑 0（低电平），要改变锁存器的输出状态，R 或 S 必须输入高电平 1。但是还要考虑一个新因素即第三个输入端。这个输入端一般标为 C 或 CLK，因为它是由某种时钟电路控制的，起到控制若干个这样的锁存器同步工作的作用，只有当 CLK 输入是逻辑 1 时输出才可能改变，当 CLK 是逻辑 0 时，S 和 R 端的输入不起作用。

要遵守不能同时在 S 和 R 输入端加信号（即同时为高电平）的规则，如果时钟端为逻辑 1 时在 S 和 R 输入端都为 1，锁存器导通，两个输出均为逻辑 1，这时如果 CLK 输入先降为逻辑 0，毫无疑问，出现一种竞争的情况，将无法确定输出端 Q 和 $\overline{Q}$ 中哪一个输出为逻辑 0，图 8.2(a)中的电路表现出了这种不确定性。

2. RS 触发器

边缘触发的 RS 触发器实际上由两个同样的 RS 锁存电路组成，RS 触发器的电路符号如图 8.2(b)所示。在两个时钟输入之间接了一个反相器，以保证这两个部分（RS 锁存电路）分别在时钟信号的不同半周期中工作（即一个在时钟信号为高电平时工作，另一个则在时钟信号为低电平时工作），这就是这个电路的关键。所以，Q 和 $\bar{Q}$ 输出只能在时钟信号从 1 降到 0 时即时钟信号的下降沿发生改变。因此就称为边缘触发的触发器。

3. JK 触发器

JK 触发器在各种情况下的输出都是完全确定的，因此对大部分逻辑电路设计来说选择 JK 触发器比较好。

一个 JK 触发器的 Q 和 $\bar{Q}$ 输出只在 CLK 信号的下降沿时改变状态，这时 J 和 K 的输入状态将决定 JK 触发器的输出状态。

如果 J 和 K 输入端（图 8.2(c)）都保持逻辑 1（高电平），而时钟信号不断地交变时，Q 和 $\bar{Q}$ 输出将在 CLK 的每个下降沿发生改变。这一特点在许多地方都很有用。一个专门以这种方式工作的触发器（J 与 K 连接在一起）用 T 表示，构成 T 触发器。

这种方式工作的 JK 触发器必须是边缘触发式的，任何高电平触发式的 JK 锁存电路如果三个输入端均为逻辑 1 时会高速振荡，这种电路是没什么用的。同样，T 触发器也必须是边缘触发式的，边缘触发是保证这两种触发器每来一个时钟脉冲仅改变一次输出状态的唯一方法。

4. 边缘触发的 D 触发器

D 触发器（图 8.2(d)）也是当时钟脉冲从逻辑 1 降到逻辑 0 时可能改变输出状态，且 Q 端的输出总是与在时钟下降沿时 D 的输入状态一致，这一点是与 RS 和 JK 触发器不同的。

8.6 阅读材料参考译文

8.6.1 74 系列集成逻辑电路

有一系列编号为 74××00 的逻辑芯片，其中间字母（××）表明芯片电路的类别，例如：74LS00 和 74HC00，最早的系列（现已不用了）中间无字母，如 7400。

74LS（低功耗肖特基——一种集成电路工艺）系列（和最早的 7400 系列一样）是用 TTL（三极管-三极管-逻辑）电路，与后面其他系列相比其速度快但功耗较大，74 系列常被称为"TTL 系列"，尽管后来的新（74 系列）芯片并不用 TTL。

74HC 系列是把 4000 系列的低功耗与 TTL 的高速相结合的高速 CMOS 电路，是与原 74LS 系列有相同引脚定义的 CMOS 芯片，要注意 74HC 系列芯片的输入是不能用 74LS 系列芯片输出驱动的，因为所用的逻辑 0 的电压范围是不匹配的。

74HCT 系列是为使 74HC 系列与 74LS 系列（TTL）输入相匹配而设计的一种特殊系列，所以 74HCT 可以与 74LS 系列在同一电路中混合使用，实际上在大多数电路中 74HCT 可以作为旧的 74LS 系列芯片的替代品。74HCT 的小缺点是抗干扰能力较差，但在大部分情况中这个缺点无关紧要。

8.6.2 寄存器

一个寄存器是一组触发器，每个触发器能存储一个二进制位的信息。一个 n 位的寄存器有 n 个触发器，能存储 n 位二进制位的信息，除了触发器，一个寄存器可能还有一些组合门电路来完成数据处理任务。按广义定义，寄存器由一组触发器和可实现转换传输的门电路组成。触发器保存二进制信息，门电路控制何时和如何把新的信息送入到寄存器中。

各种寄存器都已有成品了,最简单的寄存器是只有触发器而没有外部门电路的。图 8.3 是用 4 个 D 触发器构成的最简单的寄存器。所有的触发器在同一时钟输入脉冲的上升边缘触发,这时在 4 个输入端的二进制数据信息被传送到 4 位寄存器中。可以在任何时候读取 4 个输出端,获得存储在寄存器中的二进制信息。clear(清除)输入与每个触发器的一个特殊端(清零端)相连,当这个输入为 0 时,所有的触发器被同时清零。清零端是用来使所有的寄存器在时钟输入之前(即寄存之前)先清零。清零端平时(在时钟控制工作时)必须保持为逻辑 1,注意时钟信号是可以控制 D 的输入的,但清零输入却与时钟无关(不受时钟信号控制的)。

把新的信息传送到寄存器中被称为装载(写入)寄存器。如果寄存器的所有位是用同一个时钟脉冲控制同时写入的,我们称为并行输入。图 8.3 寄存器中,在 C 端加一个时钟控制信号将使 4 个输入($I_0 \sim I_3$)并行写入。在这种(并行)设置中,如果要保持寄存器中的内容不变的话必须禁止时钟控制端输入(脉冲)。

8.6.3 计数器

所有的计数器都需要一个"方波"时钟信号来使计数器计数。

1. 异步和同步计数器

计数器主要有两种:异步和同步,在简单的电路中,这两种计数器表现出来的性能几乎没什么差别,但它们的内部结构是不同的。

一个异步计数器中包含了若干个触发器,每个触发器的输出端与下一个触发器的输入端相连接。每当输入信号从高电平到低电平(在下降沿)时触发器的输出状态发生变化。这种简单的接法工作起来很有效,但由于时钟"异步"通过各个触发器,触发器的输出稍有延迟。

图 8.4 显示如何连接异步计数器,每个计数器的最高位输出 QD 接到下一个计数器的时钟输入端($\overline{CK}$)。

一个同步的计数器内部电路结构比较复杂,以保证所有的计数器输出都在同一个时钟脉冲输入时改变,同步的计数器可以避免异步计数器中(可能出现的)误计数。

图 8.5 给出如何连接同步计数器,所有的时钟输入端(CK)是连接在一起的,进位输出信号(CO)接在下一个计数器的进位输入信号(CI)上,第一个计数器的(CI)应接高电平(无进位信号)。

2. 上升沿和下降沿时钟输入

当时钟信号输入时开始计数,大部分同步计数器在上升沿即时钟信号从低电平到高电平跃变时计数。大部分异步计数器则在下降沿即时钟信号从高电平到低电平跃变时计数(如图 8.6 所示)。

3. 74××390 双十进制异步计数器

74××390 中含有两个独立的十进制计数器(如图 8.7 所示),芯片两边各有一个。它们是异步计数器,所以要注意由于后面的计数器的输出相对于时钟脉冲有一个很小的延迟,可能会在与它们相连的各个逻辑门系统中出现干扰信号。

当时钟信号跃变为 0(在下降沿)时开始计数。通过在时钟输入端上加一横线表示是下降沿有效。异步计数器的时钟信号通常都是这样(下降沿有效),所以异步计数器的输出信号可以直接驱动下一个(异步)计数器的时钟输入。

每个计数器由两个部分组成:时钟 clockA(控制)-QA(二进制计数器)和时钟 clockB

（控制）-QB-QC-QD（五进制计数器），一般的用法是将 QA 与时钟 clockB 连接，从而把两部分连接起来（构成十进制计数器），外部时钟信号接在时钟 clockA 端。

8.6.4　7 段显示驱动（芯片）

显示驱动器的输入端 A～D 与十进制计数器输出端 QA～D 相连接（如图 8.8 所示），内部的逻辑门电路使显示驱动器的 a～g 输出端输出高电平或低电平，正好使一个 7 段显示器的对应的 a～g 段点亮（显示一个数字）。在每个输出端都要串联一个电阻以保护二极管，对电源电压为 4.5～6 V 的 LED 数码管电阻取 330 Ω较合适。注意有的电路图上略去了这些电阻（但实际中必须接电阻）。

有两种 7 段 LED 显示器：

共阳极（CA 或 SA），其所有的二极管的阳极或称正极连接在一起，这个公共阳极接+U_S。这种显示器需要接共阳显示驱动芯片，如 74××47，它能输出低电平使对应的二极管发光。

共阴极（CC 或 SC），其所有的二极管的阴极连接在一起，这个公共阴极接 0 V。这种显示器需要接共阴显示驱动芯片，如 4511 等，它能输出高电平使对应的发光二极管发光。

LED 显示器通常有两个共阳或共阴引脚，显示器还有一个小数点（DP）引脚，但它不受显示驱动芯片控制，较大的显示器还有用两个发光二极管串联的（使显示的数码比较大）。

Unit 9　Microcomputers

Pre-reading

Read the following passage, paying attention to the question.
1) What make up a computer system?
2) What is the system bus?
3) What is the function of the main memory?

9.1　Text

9.1.1　Basic Computer

The computer that everyone thinks of first is typically the personal computer, or PC. The basic components that make up a computer system: the CPU, memory, I/O, and the bus that connects them. Although you can write software that is ignorant of these concepts, high performance software requires a complete understanding of this material.

The basic operational design of a computer system is called its architecture. John Von Neumann, a pioneer in computer design, gave the architecture of most computers in use today. A typical Von Neumann system has three major components: the central processing unit (or CPU), memory, and input/output (or I/O). In a system, the way to these components design impacts system performance. In VNA machines, like the 80x86 family, the CPU is where all the action takes place. All computations occur inside the CPU. Data and CPU instructions reside in memory until required by the CPU. To the CPU, most I/O devices look like memory because the CPU can store data to an output device and read data from an input device. The major difference between memory and I/O locations is the fact that I/O locations are generally associated with external devices in the outside world.

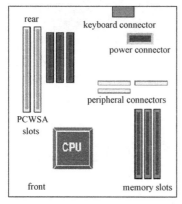
Fig 9.1　the motherboard

9.1.2　the Motherboard

The motherboard is the main circuit board inside the PC which holds the processor, memory and expansion slots and connects directly or indirectly to every part of the PC (Fig 9.1). It's made up of a chipset (known as the "glue logic"), some code in ROM and the various interconnections or buses. PC designs today use many different buses to link their various components. Wide, high-speed buses are difficult and expensive to produce, the signals travel at such a rate that even distances of just a few centimeters cause timing problems; while the metal tracks on the circuit board act as miniature radio antennae, transmitting electromagnetic noise that introduces interference with signals elsewhere in the system. For these reasons, PC design engineers try to keep the fastest buses confined to the smallest area of the motherboard and use slower, more robust buses, for other parts.

9.1.3 the System Bus

The system bus connects the various components of a VNA machine. The 80x86 family has three major busses: the address bus, the data bus, and the control bus. A bus is a collection of wires on which electrical signals pass between components in the system. These busses vary from processor to processor. However, each bus carries comparable information on all processors, for example, the data bus may have a different implementation on the 80386 than on the 8086, but both carry data between the processor, I/O, and memory.

1. the Data Bus

The 80x86 processors use the data bus to transfer data between the various components in a computer system. The size of this bus varies widely in the 80x86 family. Indeed, this bus defines the "size" of the processor.

2. the Address Bus

The data bus on an 80x86 family processor transfers information between a particular memory location or I/O device and the CPU. The only question is, "Which memory location or I/O device?" The address bus answers this question.

To differentiate memory locations and I/O devices, the system designer assigns a unique memory address to each memory element and I/O device. When the software wants to access some particular memory location or I/O device, it places the corresponding address on the address bus. Circuitry associated with the memory or I/O recognizes this address and instructs the memory or I/O device to read the data from or place data on the data bus.

3. the Control Bus

The control bus is a collection of signals that control how the processor communicates with the rest of the system. Consider for a moment the data bus. The CPU sends data to memory and receives data from memory on the data bus. This prompts the question." Is it sending or receiving?" There are two lines on the control bus, read and write, which specify the direction of data flow. Other signals include system clocks, interrupt lines, status lines, and so on. The exact make up of the control bus varies among processors in the 80x86 families. However, some control lines are common to all processors and are worth a brief mention.

The read and write control the direction of data on the data bus. When both contain logic one, the CPU and memory, I/O is not communication with one another. If the read line is low (logic zero), the CPU is reading data from memory (that is, the system is transferring data from memory to the CPU). If the write line is low, the system transfers data from the CPU to memory.

9.1.4 Main Memory

The main memory is the central storage unit in a computer system. It is a relatively large and fast memory used to store programs and data during the computer operation. The principal technology used for the main memory is based on semiconductor integrated circuits. Integrated circuits RAM (Random Access Memory) chips are available in two possible operation modes, static and dynamic. The static RAM consists essentially of internal flip-flops that store the binary information. The stored information remains valid as long as power is applied to the unit. The dynamic RAM stores the binary information in the form of electric charges that are applied to the capacitors. The capacitors are provided inside the chip by MOS transistors. The stored charges on the capacitors tend to discharge with time and refreshing the dynamic memory must periodically recharge the capacitors. The dynamic RAM offers reduced power consumption and large storage capacity in a single memory chip. The static RAM is easier to use and has shorter read and write cycles.

9.1.5 BIOS (Basic Input/Output System)

1. Introduce to BIOS

One of the most common uses of Flash memory is for the basic input/output system of your computer, commonly known as the BIOS (Fig 9.2). The BIOS makes sure all the other chips, hard drives, ports and CPU function together.

Fig 9.2   BIOS uses Flash memory

The BIOS is special software that interfaces the major hardware components of your computer with the operating system. It is usually stored on a Flash memory chip on the motherboard, but sometimes the chip is another type of ROM (Read Only Memory).

2. What BIOS Does

The BIOS software has a number of different roles, but its most important role is to load the operating system. When you turn on your computer and the microprocessor tries to execute its first instruction, it has to get that instruction from somewhere. It cannot get it from the operating system because the operating system is located on a hard disk, and the microprocessor cannot get to it without some instructions that tell it how. The BIOS provides those instructions. Some of the other common tasks that the BIOS performs include:

A power-on self-test (POST) for all of the different hardware components in the system to make sure everything is working properly.

Activating other BIOS chips on different cards installed in the computer - For example, SCSI and graphics cards often have their own BIOS chips.

Providing a set of low-level routines that the operating system uses to interface to different hardware devices - It is these routines that give the BIOS its name. They manage things like the keyboard, the screen, and the serial and parallel ports, especially when the computer is booting.

The first thing the BIOS does is check the information stored in a tiny (64 bytes) amount of RAM located on a CMOS chip. The CMOS setup provides detailed information particular to your system and can be altered as your system changes. The BIOS uses this information to modify or supplement its default programming as needed.

Interrupt handlers are small pieces of software that act as translators between the hardware components and the operating system. For example, when you press a key on your keyboard, the signal is sent to the keyboard interrupt handler, which tells the CPU what it is and passes it on to the operating system. The device drivers are other pieces of software that identify the base hardware components such as keyboard, mouse, hard drive and floppy drive. Since the BIOS is constantly intercepting signals to and from the hardware, it is usually copied, or shadowed, into RAM to run faster.

Technical Words and Phrases

address	[ə'dres]	vt. 访问（写地址，向……提议） n. 地址（姓名，住址）
architecture	[ˈaːkitektʃə(r)]	n. 计算机的物理结构，包括组织结构、容量、该计算机的 CPU、存储器以及输入输出设备间的互联
assign	[ə'saɪn]	vt.（与 to 连用）分配；指定（把时间、地点等）
bus	[bʌs]	n.（计算机）总线
cycle	['saɪk(ə)l]	n. 循环；周而复始，周期；循环期
differentiate	[ˌdɪfəˈrenʃɪeɪt]	vt. 区分，区别，辨别
dynamic	[ˌdaɪˈnæmɪk]	adj. 动力的，动态的
family	[ˈfæmɪlɪ]	n. 系列，家族 adj. 家庭的，家族的
ignorant	[ˈɪɡnərənt]	adj.（常与 of, in 连用）无知识的；不知道的
location	[ləʊˈkeɪʃ(ə)n]	n. 地点，位置
motherboard	[ˈmʌðəbɔːd]	n.（计算机）主板
periodically	[ˌpɪərɪˈɒdɪkəːlɪ]	adj. 周期的，期刊的 n. 期刊，杂志
prompt	[prɒmpt]	vt.（常与 to 连用）提出，促使；提示 n. 提示，提词，提示符 adj. 迅速的；及时的
refresh	[rɪˈfreʃ]	vt. 刷新，消除疲劳；恢复精神
size	[ˈsaɪz]	vt. 计算机可同时处理数据的二进制位数
communicate with...		与……联络；通信；交换（看法等）
give sb credit for sth		为……赞扬某人，认为某人具有（某种品德、才能等）

Notes to the Text

1. **John Von Neumann** 冯·诺依曼，第一个提出计算机工作原理为先存储、后执行的概念，现在绝大部分计算机都按此原理构成，故又把按这原理构成的计算机系统称为 Von Neumann system 或 VNA machine（冯·诺依曼结构）。

2. the 80x86 family　Intel 公司开发的 8086,80286,80386,80486 等系列计算机 CPU 芯片。

9.2 Reading Materials

9.2.1 Microcontroller

Intel's 8-bit MCS-51 family of microcontrollers is a leading choice for embedded control. This Classic family consists of CHMOS versions of all the original 8-bit microcontrollers that introduced the MCS-51 microcontroller family. Intel offers a wide variety of on-board memory in EEPROM (89C51), EPROM (87C51), ROM (80C51), as well as CPU-only (80C31) microcontrollers. Thus, the ability to program an 8051 is an important skill for anyone who plans to develop products that will take advantage of microcontrollers.

The 8051 has three very general types of memory. To effectively program the 8051 it is necessary to have a basic understanding of these memory types.

The memory types are illustrated in the graphic (Fig 9.3). They are: On-Chip Memory, External Code Memory, and External RAM.

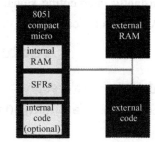

On-Chip Memory refers to any memory (Code, RAM, or other) that physically exists on the microcontroller itself.

Fig 9.3　the memory types of 8051

On-chip memory can be of several types. External Code Memory is code (or program) memory that resides off-chip. This is often in the form of an external EPROM. External RAM is RAM memory that resides off-chip. This is often in the form of standard static RAM or flash RAM.

9.2.2　about DNA Computers

You won't believe where scientists have found the new material they need to build the next generation of microprocessors. Millions of natural supercomputers exist inside living organisms, including your body. DNA (deoxyribonucleic acid) molecules, the material our genes are made of, have the potential to perform calculations many times faster than the world's most powerful human-built computers. DNA might one day be integrated into a computer chip to create a so-called biochip that will push computers even faster.

Leonard Adleman, a computer scientist at the University of Southern California, is often called the inventor of DNA computers. His article in a 1994 issue of the journal Science outlined how to use DNA to solve a well-known mathematical problem, called the directed Hamilton Path problem, also known as the "traveling salesman" problem. The goal of the problem is to find the shortest route between a number of cities, going through each city only once. As you add more cities to the problem, the problem becomes more difficult. Adleman choses to find the shortest route between seven cities using his DNA test-tube computer.

The success of the Adleman DNA computer proves that DNA can be used to calculate complex mathematical problems. However, this early DNA computer is far from challenging silicon-based computers in terms of speed. Another drawback of his DNA computer is that it requires human assistance. The goal of the DNA computing field is to create a device that can work independent of human involvement.

Three years after Adleman's experiment, researchers at the University of Rochester developed logic gates made of DNA. These DNA logic gates are the first step toward creating a computer that has a structure similar to that of an electronic PC. Instead of using electrical signals to perform logical operations, these DNA logic gates rely on DNA code. The researchers believe that these logic gates might be combined with DNA microchips to create a breakthrough in DNA computing.

DNA computer components - logic gates and biochips - will take years to develop into a practical, workable DNA computer. If such a computer is ever built, scientists say that it will be more compact, accurate and efficient than conventional computers.

9.2.3 PLC

1. General

Programmable logic controllers (PLC) have been around recently (Fig 9.4). Their proven reliability in harsh environments and design to handle many inputs and outputs has made them the foundation of many factory automated systems.

The basic components of a PLC are the rack, power supply, CPU, and I/O modules.

Most PLCs have a separate rack and power supply to make them modular. For small systems we use a single small rack and power supply and for large systems we use multiple large racks and power supplies. Note that the power supply is intended to power the cards that plug into the rack and not for powering the sensors and actuators. The power for sensors and actuators should be powered using a separate, isolated power source (AC power or preferably a separate 24 V DC power supply).

Fig 9.4   PLC

2. Brick PLCs

The smallest PLCs sometimes have the CPU, rack, power supply and I/O all combined into a "brick" configuration. They are called "brick PLCs" because they are about the size of a brick. Manufacturers do this to reduce the total cost of the PLC as the cost of these "brick" PLCs typical cost between $100 and $500. These PLCs do make sense in very simple and limited applications that will never be expanded in the future. Sometimes we would rather spend a little more money and have a modular system with future expansion capabilities.

3. PLC and Computer

Sometimes we do not use a typical PLC for the control engine. Typically what we do is:

If the application is small (less than 50 I/O), no databases (only a few choices), then use a PLC.

If the application is small, slow (response time greater than 50 milliseconds) and requires computer functionality (machine vision, networking, databases, multiple axis motion control, etc) we prefer to do the entire application in Visual Basic (VB) or C#.

If there are large amounts of I/O (over 100) or you need fast, real-time response, then you will probably want both - the PLC handling your real-time and direct I/O tasks and the computer handling the non-real-time tasks (such as databases, etc).

9.3 Knowledge about Translation
（翻译知识 9——连词Ⅰ）

连词是用来连接词与词，短语与短语或句子与句子的词。根据连词的作用，连词分为并列连词和从属连词。并列连词连接并列的词、短语和句子，如 and, but 和 or 等；从属连词连接主句和从句，如 that, which, when 等。从属连词只起到连接句子的作用，在从句中不起成分作用。在科技文章中常见到连词，通过连词能正确断开句子，把复杂长句分为较简单的句子便于理解。这里主要介绍从属连词。

1. who, whom, whose

who (whom)可用做关系代词，引导定语从句，在分析时把从句找出，句子翻译就简单了。

Scientists **who** deal with the physical universe must deal with both matter and energy. 研究物质的世界的科学家，不仅要研究物质，而且要研究能量。

whose 可用来引导定语从句，它既可以代替人，也可以代替物，既可以代替单数名词，也可代替复数名词。

Perhaps light is some sort of electric wave **whose** nature we do not yet understand. 或许，光是某种电磁波，它的性质我们还不十分清楚。

2. while

while 可作为连接时间状语从句的连词，译成：（正）当……的时候。

while 可作为连接让步状语从句的连词，译成：虽然，尽管。

while 还可作为连接两个并列句的连词，译成：而，却，可是。

Metals in general are good conductors, **while** nonmetals are insulators. 通常，金属是良导体，而非金属是绝缘体。

3. when

① when 引导时间状语从句：

When we talk of electric current, we mean electrons in motion. 当我们谈论电流时，我们指的是运动的电子。

when＋分词、介词或形容词，这种结构可以看成是省略句，只有从句中的主语和主句中的主语一致而且从句中的谓语一部分是用 be 来表示时，才可以构成这种句型。

② when 引导定语从句，用来修饰表示时间概念的名词：

There are times **when** it is advantageous to substitute one kind of gate for another. 有时用一种门去替代其他门更有利。

4. after, before, until

after（在……之后），before（在……之前），until (till) 可用做连词，连接时间状语从句。尤其是 until(till)，翻译时注意，若主句是肯定句，可译成"直到……为止"。

CPU instructions reside in memory **until** required by the CPU. CPU 指令存储在存储器中直到 CPU 需要时（才调出）。

若主句是否定句，则译成"直到……才"，"在……之前不……"。

The content of the register will be left unchanged **until** a clock transition applied to the C input of the register. 直到在寄存器的 C 输入端加上一个时钟（脉冲）跃变，寄存器中的内容才会改变。

5. where

where 在科技英语中常引导地点状语从句和定语从句，where 也可以引导限定性和非限定性定语从句。

Where the ohm is too small a unit, we may employ the kilo-ohm and megohm. 如果欧姆这个单位太小，则可用千欧和兆欧。这里的 where 指的不是具体的地点，可译成"如果在……情况下"，"如果在……条件下"。

6. how, why

科技英语中常用 how 来引导描述操作过程的从句，而 why 则常用来引导说明原因的从句。

The first thing we have to know is **how** bodies become charged with electricity. 我们必须了解的第一件事，就是物体是怎样带电的。

The question is **why** metals are good conductors. 问题在于为什么金属是良导体。

7. because, since, for

这三个词用做连词时所引导的句子都可以表示原因。because 表示事物本质的原因，直接的理由，事物内在的必然的因果关系；since 只是表示事物内在联系上一种合乎逻辑的自然结果；for 表示间接的、附加的理由，或者是一种推断的理由，并非本质的原因。这三个词，because 语气最强，since 其次，而 for 是并列连词，连接两个并列分句。

Since the BIOS is constantly intercepting signals to and from the hardware, it is usually copied, or shadowed, into RAM to run faster. 因为 BIOS 是不断地从硬件接收信号和输出信号给硬件的，所以为了运行更快，（计算机启动后）一般把 BIOS 复制到 RAM 中。

9.4 Exercises

1. Put the Phrases into English

① 组成一个计算机系统　　　　　⑥ 控制信号线
② 高性能的软件　　　　　　　　⑦ 中断线
③ 输入输出器件　　　　　　　　⑧ 从存储器中读数据
④ 存储在存储器中的 CPU 指令　 ⑨ 半导体集成电路
⑤ 把数据存储到一个输出设备中　⑩ 动态随机存储器

2. Put the Phrases into Chinese

① impact system performance
② be generally associated with external devices
③ read data from an input device

④ connect the various components of a VNA machine
⑤ data bus
⑥ 80x86 family microprocessor
⑦ to differentiate memory locations and I/O devices
⑧ this prompts the question
⑨ worth a brief mention
⑩ storage capacity in a single memory chip

3. Sentence Translation

① To the CPU, most I/O devices look like memory because the CPU can store data to an output device and read data from an input device.

② A bus is a collection of wires on which electrical signals pass between components in the system.

③ The signals travel at such a rate that even distances of just a few centimeters cause timing problems.

④ When the software wants to access some particular memory location or I/O device, it places the corresponding address on the address bus.

⑤ The main memory is a relatively large and fast memory used to store programs and data during the computer operation.

⑥ The stored information remains valid as long as power is applied to the unit.

⑦ In VNA machines, like the 80x86 family, the CPU is where all the action takes place.

⑧ The static RAM is easier to use and has shorter read and write cycles.

⑨ One of the most common uses of Flash memory is for the basic input/output system of your computer, commonly known as the BIOS.

⑩ The BIOS software has a number of different roles, but its most important role is to load the operating system.

4. Translation

① The credit goes to the British engineer John Logic Baird **who** followed the footprints of Marconi and tried to send the images in the same way as the speech.

② An inductor is an electrical device, **which** can temporarily store electromagnetic energy in the field about it.

③ The control bus is a collection of signals that control **how** the processor communicates with the rest of the system.

④ People often prefer analog oscilloscopes **when** it is important to display rapidly varying signals in "real time".

⑤ These heavy installation costs are required **because** robots tend to be insensitive to their surroundings.

9.5 课文参考译文

9.5.1 基本型计算机

一提到计算机，人们首先想到的是个人计算机，简称 PC。组成一个计算机系统的基本部分是：中央处理器 CPU、存储器、输入/输出端口和把它们连接在一起的总线。虽然你在编写软件程序时可以不考虑这些概念，但编写高性能的软件却需要对这些部件有一个完整的了解。

对一个计算机系统的基本设计被称为计算机的结构模型，冯·诺依曼，一个计算机设计的先驱，提出了今天所用的大部分计算机的结构模型。一个典型的冯·诺依曼系统有三个主要的部件：中央处理器（即 CPU）、存储器及输入/输出口。一个系统中，设计这些部件的方法影响了系统的性能。在冯·诺依曼结构中，如 80x86 系列计算机，所有的操作都在中央处理器中执行。所有的计算都在中央处理器内部进行，数据和指令存储在存储器，由中央处理器调用。对 CPU 而言，大量的输入/输出口就像存储器一样，因为 CPU 可以把数据存储在输出设备中，也可从输入设备中读入数据。存储器和输入/输出口主要的区别是输入/输出口通常是与（外部世界的）外部器件相连接的。

9.5.2 主板

主板是在计算机内部的主电路板，上面有中央处理器（CPU）、存储器和扩展槽，并与计算机的各个部分直接或非直接地相连（如图 9.1 所示）。主板由芯片组（称为 glue logic）、一些 ROM 中的代码和各种接口或总线组成。现在计算机中用各种不同的总线连接各种各样的部件。高带宽的、高速总线生产麻烦且成本高，（因）其信号的传输速度非常快，信号线只要有几厘米的传输距离就可能引起时间（延迟）问题；而电路板上的金属导线像微型无线电天线，会发射电磁噪声，从而对系统中的信号产生干扰。因此，计算机（主板）设计者要把最高速的总线限制在主板的（某块）最小的面积中，而把低速的、较粗的总线放在其他地方。

9.5.3 系统总线

系统总线连接冯·诺依曼结构中的各个部件，80x86 系列计算机有三种主要的总线：地址线、数据线和控制线。总线是指一组在系统各个部件之间传递各种电信号的导线。对各种处理器所需要的总线不同。但是，对所有的处理器而言，每种总线都携带相应的信息，如 80386 和 8086 有不同的数据总线（数目），但都是在处理器、输入/输出口及存储器之间传递数据信息。

1. 数据总线

80x86 系列处理器用数据总线在一个计算机系统的各种部分之间传递数据。在 80x86 系列中这组总线的数目是不同的，实际上，它决定了处理器的"大小"（即多少位数据的处理器）。

2. 地址总线

80x86 系列处理器中数据总线在中央处理器和一个特定的存储器位置或输入/输出口之间传递信息，但一个问题是：如何确定存储器或输入/输出口的位置？地址总线回答了这个问题。

对于不同的存储位置和输入/输出设备，系统的设计者指定一个唯一的存储地址。当软件想要访问一些指定的存储位置或输入/输出设备时，它把相应的地址值放在地址总线上，与存储器或输入/输出设备相连接的电路识别出这个地址，命令存储器或输入/输出设备从数据总线上读取数据或输出数据到数据总线上。

3. 控制总线

控制总线是一组控制处理器如何与系统的其他部分通信的信号线的集合。再来看数据总线，CPU 是通过数据总线把数据送到存储器中和从存储器中接收数据的，这就提出了一个问题"是送数据还是接收数据？"控制总线中有两根线：读和写，就是用来指定数据流动的方向的。其他的信号线包括系统时钟线、中断线、状态线等。在 80x86 系列处理器中控制总线的数目是各不相同的，然而，有些控制线是所有的处理器都有的，值得作简短介绍。

读和写控制线控制数据总线上数据流的方向，当这两根线都为逻辑 1 时，CPU 和存储器、输入/输出口互相之间是不通信的；如果读控制线是低电平（逻辑 0），则 CPU 从存储器中读取数据（即系统是把数据从存储器中传送到 CPU 中的），如果写控制线是低电平的，系统把数据从 CPU 传送到存储器中。

9.5.4 主存（内存）

主存是一个计算机系统中的中心存储单元。它是一个相当大且存取速度很快的存储器，用来存储 CPU 操作时的程序和数据。主存主要采用半导体集成电路技术。有两种可用的集成 RAM 芯片，即静态 RAM 和动态 RAM。静态 RAM 本质上是用内部的触发器组成的，用来存储二进制信息。只要芯片与电源相连，静态 RAM 存储的信息始终有效。动态的 RAM 以电容两极（充电）电荷的形式存储信息，芯片内的电容是由 MOS 管构成的。电容两端存储的电荷会随时间的延长而放电，要保持动态存储信息必须周期性地对电容再充电（刷新）。动态 RAM 功耗小，单一芯片的存储量大。静态 RAM 用起来方便且有较短的读和写周期（即读写速度快）。

9.5.5 BIOS（基本输入/输出系统）

1. BIOS 简介

闪存最主要的功能之一是用于计算机的基本输入/输出系统（通常称做 BIOS）（如图 9.2 所示）。BIOS 使得所有其他芯片、硬件、端口和 CPU 一起工作。

BIOS 是一种特殊的软件，是主要硬件与操作系统的接口软件，通常存储在主板的闪存芯片中，但有时也可能存储在一种 ROM 芯片中。

2. BIOS 的作用是什么

BIOS 软件有很多不同的作用，但它最重要的作用是载入操作系统。当你开机时，微处理器开始执行第一条指令，它必须从某个地方得到这条指令（就是存储在闪存中的 BIOS）。微处理器不能执行操作系统，因为操作系统存储在硬盘上，没有指令 CPU 不能从硬盘上载入操作系统，BIOS 提供了这些指令。BIOS 还提供了一些其他指令(程序)，如：

电源自检（POST）用来检查系统中各个硬件是否工作正常。

激活（调用）计算机中其他不同卡上 BIOS 芯片中的程序，如 SCSI（接口卡）和图像处理卡本身自带 BIOS 芯片。

提供操作系统用于连接不同硬件接口的一组低级处理程序，正是这些低级程序赋予了 BIOS 一词的含义。尤其是计算机启动时这些程序管理键盘、显示屏、串行和并行接口等设备。

BIOS 所做的第一件事是检查存储在一微小（64 字节）RAM（一个 CMOS 芯片）中的信息，CMOS 给出了你系统的设置信息，当你系统改变时可以改变 CMOS 设置，BIOS 要根据这些信息调整或补充它的默认程序。

中断处理是一小段软件程序，就像是硬件与操作系统之间的翻译。例如当你在键盘上按下一个键，信号就调用了键盘的中断处理程序，告诉 CPU 是什么键并发送给操作系统。设备驱动是另外的软件程序，它识别基本的硬件，如键盘、鼠标、硬盘和软盘驱动器。因为 BIOS 是不断地从硬件接收信号和输出信号给硬件的，所以为了运行更快，（计算机启动后）一般把 BIOS 复制到 RAM 中。

9.6 阅读材料参考译文

9.6.1 单片机（微控制器）

英特尔公司的 8 位 MCS-51 系列单片机是嵌入式控制（系统）的首选。这个经典系列包含了全部原 MCS-51 系列 8 位单片机的 CHMOS 版本。英特尔提供了很多种选择：片内含有 EEPROM（89C51），EPROM（87C51），ROM（80C51）存储器，以及片内不含存储器（80C31）的单片机。因此对一个想要用微处理器开发产品的人来说掌握 8051 编程技术是很重要的。

8051 有 3 种存储器，为了掌握 8051 编程（技术）需要了解这些基本的存储类型。

图 9.3 的图片说明这三种存储器，它们是片内存储器、外部程序存储器和外部数据存储器。

片内存储器是指单片机芯片内自带的存储器（程序、数据和其他类型），片内存储器可以有多种形式。外部程序存储器是代码（程序）存储器，一般是外接的 EPROM 芯片。外部 RAM 存储器也是外部芯片，通常是静态 RAM 或闪存 RAM 芯片。

9.6.2 DNA 计算机

你可能不会相信，科学家已经找到制作新一代微处理器的材料了。其实在生物体包括你的身体中就有数百万个天然的超级计算机，它就是组成我们的基因的 DNA（脱氧核糖核酸）分子，其运算速度有可能比人类制造的世界上功能最强的（电子）计算机要快很多倍。也许在某一天，可把 DNA 集成在计算机芯片中制成一个生物芯片，将使计算机运行得更快。

Leonard Adleman，美国南加利福尼亚大学的一位计算机科学家，常被称为 DNA 计算机的发明者，他于 1994 年在一个科学期刊上发表了一篇文章，提出如何用 DNA 去求解一个众所周知的被称为哈密尔顿路径（又称推销员旅行）的数学问题，问题的目标是找到一条走遍若干个城市，每个城市仅去一次且必须去一次的最短路径。当城市越来越多时，这个问题就越来越复杂。Adleman 用他的 DNA 试验计算机求解了 7 个城市的最短路径。

Adleman DNA 计算机的成功证明了 DNA 可以用来计算复杂的数学问题，但是这个早期的 DNA 计算机在计算速度上是远不如（用硅制造的电子）计算机的。他的 DNA 计算机的另一个缺点是需要人的协助。DNA 计算领域的最终目标是造出可以不需要人干预的能独立工作的器件。

在 Adleman 实验后的第 3 年，罗切斯特大学的研究人员开发出 DNA 逻辑芯片。这些 DNA 逻辑门是制作出类似于电子计算机结构的 DNA 计算机的第一步。与用电信号进行逻辑运算不同的是，这些 DNA 逻辑门用 DNA 代码进行逻辑运算。研究人员相信这些逻辑门可以组合成 DNA 芯片实现 DNA 运算中的新突破。

DNA 计算机部件——逻辑门和生物芯片，数年后将可能发展成一个实用的 DNA 计算机。如果这个计算机能制造出来的话，科学家认为它将比传统计算机更加微小、精确和高效。

9.6.3　PLC（可编程逻辑控制器）

1. 概述

可编程逻辑控制器（PLC）近年来到处可见（如图 9.4 所示），PLC 在恶劣工作环境中工作的可靠性和可以处理许多输入/输出节点的设计使它们成为许多工厂自动化系统的基础（设备）。

PLC 的主要部件是插件箱、电源模块、CPU 模块和 I/O 模块。

许多 PLC 有分开的插件箱和电源，可以（根据需要）组合，对小系统可用一个小插件箱和一个电源，对大的系统我们用多个大插件箱和多个电源，注意电源只能对插件箱中的卡供电，并不能对传感器和执行机构（如电动机）供电。传感器和执行器要由另外的隔离电源供电（交流电源或独立 24 V 直流电源）。

2. 一体式 PLC

微小型 PLC 有时把 CPU、插件箱、电源和输入/输出接口组合成一体，称为砖式 PLC（文中译做一体式），因为其大小像一块砖。制造商为了减小总成本，生产这种一体式 PLC，一体式 PLC 成本仅在 100～500 美元之间。这种 PLC 很简单且使用范围也有限（以后永远不能扩展接口）。有时我们宁可多花一些钱购买一个将来可以扩展的模块化系统。

3. PLC 与计算机

有时用 PLC 做控制设备并不一定合适，通常的选择是：

如果要控制的设备是少量的，少于 50 个 I/O 接口，不需要数据库（只有一些控制选择）就用一个 PLC。

如果需要控制设备是少量的、慢速的（响应时间大于 50 ms）并需要计算机功能（可视化、网络、数据库、多轴运动控制等），我们宁可用 VB 或 C#实现完整的应用。

如果有很多 I/O（超过 100 个）接口，或需要快速、实时响应，则 PLC 与计算机两个都需要，PLC 处理实时响应和直接输入/输出任务，计算机处理非实时任务（如数据库等）。

Unit 10 Programming the Computer

 Pre-reading

Read the following passage, paying attention to the question.
1) What is C Language?
2) What is an Assembly Language?
3) Do you know any other programming language?

10.1 Text

10.1.1 C as a Structured Language

C is commonly considered to be a structured language with some similarities to Algol and Pascal. Although the term block-structured language does not strictly apply to C in an academic sense, C is informally part of that language group. The distinguishing feature of block-structured language is that the compartmentalization of code and data. This means that a language can section off and hide, from the rest of the program, all information and instructions that are necessary to perform a specific task. Generally, compartmentalization is achieved by subroutines with local, or temporary variables. In this ways, you can write subroutines so that the events that occur within them will cause no side effects in other parts of the program. In contrast excessive use of global variables, which are known throughout the entire program, may allow bugs, or unwanted side effects, to creep into a program. In C, all subroutines are discrete functions.

Functions are the building blocks of C, in which all program activity occurs. They allow you to define and code specific tasks in a program separately. After debugging a function that uses only local variables, you can rely on it to work properly in various situations without creating side effects in other parts of your program. All variables that are declared in that function will be known only to that function.

In C, using blocks of code also creates program structure. A block of code is a logically connected group of program statements that can be treated as a unit. You create a block of code by placing lines of code between opening and closing curly braces.

```
if(x<10)
  {
    printf ("too low, try again");
    reset_counter(-1);
  }
```

The two statements after "if" that are between the curly braces are executed if x is less than 10. These two statements with the braces represent a block of code. They are linked together: one of the statements cannot execute without the other also executing. In C, every statement can be either a single statement or a block of statements. The use of code block creates readable programs with logic that is easy to follow.

C is a programmer's language. Unlike most high-level computer languages, C imposes few restrictions on what you can do with it. By using C, a programmer can avoid the use of assembly code in all but most demanding situations. In fact, one reason for the invention of C was to provide an alternative to assembly language programming.

10.1.2 MATLAB Language

MATLAB is a numerical computing environment and programming language, Maintained by the MathWorks.The MATLAB integrates computation, visualization, and programming in an easy-to-use environment where problems and solutions are expressed in familiar mathematical notation.

MATLAB is an interactive system whose basic data element is an array. It allows you to solve many technical computing problems, especially those with matrix and vector formulations. An additional package, Simulink, adds graphical multidomain simulation and Model-Based Design for dynamic and embedded systems.

MATLAB is easy to use, here are two examples:

1.Start & Quit MATLAB

When you start MATLAB, the desktop appears, containing tools (graphical user interfaces) for managing files, variables, and applications associated with MATLAB.

Fig 10.1 shows the default desktop. You can customize the arrangement of tools and documents to suit your needs.

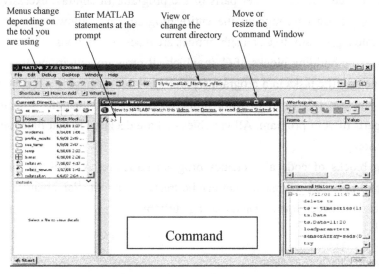

Fig 10.1   desktop of MATLAB

To end your MATLAB session, select File > Exit MATLAB in the desktop, or type quit in the Command Window.

2. Plotting Process

The MATLAB environment provides a wide variety of techniques to display data graphically. Interactive tools enable you to manipulate graphs to achieve results that reveal the most information about your data.

For example, the following statement creates a variable x that contains values ranging from -1 to 1 in increments of 0.1. The second statement raises each value in x to the third power and stores these values in y:

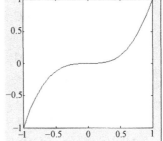

x = -1:.1:1;   % define x array
y = x.^3;   % Δ find the third power of each value in x and store as y
plot(x,y);   % draw a curve of y-x

Fig 10.2 A simple line graph

A simple line graph(Fig 10.2) is a suitable way to display x as the independent variable and y as the dependent variable.

You can also annotate and print graphs for presentations, or export graphs to standard graphics formats for presentation in Web browsers or other media.

10.1.3 Assembly Language Instructions

Assembly language instructions are provided to describe each of the basic operation that can be performed by a microprocessor. They are written using alphanumeric symbols instead of the microprocessor's machine code. The assembly language statements are located on the left. Frequently, comments describing the statements are included on the right. This type of documentation makes it easier for programmers to write, read, and debug code. By the term "code" we mean programs written in the language of the microprocessor. Programs written in assembly language are called source code.

Each instruction in source program corresponds to one assembly language statement. The statement must specify which operation is to be performed and what data operands are to be processed. For this reason, an instruction can be divided into two separate parts: its opcode and its operands. The opcode is the part of the instruction that identifies the operation that is to be performed. For example, typical operations are add, subtract, and move.

In assembly language, we assign a unique one, two, or three-letter combination to each operation. This letter combination is referred to as a mnemonic for the instruction. For instance, the 8086 assembly language mnemonics for add, subtract, and move are ADD, SUB, and MOV, respectively.

Operands identify the data that are to be processed by the microprocessor as it carried out the operation specified by the opcode. For instance, in an instruction that adds the contents of the base register to the accumulator, BX and AX are the operands. An assembly language description of this instruction is:

ADD   AX, BX

80x86 instructions can be (roughly) divided into eight different classes:
① Data movement instructions

MOV, LEA, LES , PUSH, POP, PUSHF, POPF

② Conversions

CBW, CWD, XLAT

③ Arithmetic instructions

ADD, INC SUB, DEC, CMP, NEG, MUL, IMUL, DIV, IDIV

④ Logical, shift, rotate, and bit instructions

AND, OR, XOR, NOT, SHL, SHR, RCL, RCR

⑤ I/O instructions

IN, OUT

⑥ String instructions

MOVS, STOS, LODS

⑦ Program flow control instructions

JMP, CALL, RET, CONDITIONAL JUMPS

⑧ Miscellaneous instructions.

CLC, STC, CMC

10.1.4 Introduce to operating systems

Every desktop and laptop computer in common use today contains a microprocessor as its central processing unit. The microprocessor is the hardware component. To get its work done, the microprocessor executes a set of instructions known as software. There are two different types of software:

The operating system - The operating system provides a set of services for the applications running on your computer, and it also provides the fundamental user interface for your computer.

The applications - Applications are pieces of software that are programmed to perform specific tasks. On your computer right now you probably have a browser application, a word processing application, an E-mail application and so on. You can also buy new applications and install them.

More recently, operating systems have started to pop up in smaller computers as well. The computers used in these little devices have gotten so powerful that they can now actually run an operating system and applications. The computer in a typical modern cell phone is now more powerful than a desktop computer from 20 years ago, so this progression makes sense and is a natural development. In any device that has an operating system, there's usually a way to make changes to how the device works. This is far from a happy accident; one of the reasons operating systems are made out of portable code rather than permanent physical circuits is so that they can be changed or modified without having to scrap the whole device.

For a desktop computer user, this means you can add a new security update, system patch, new application or often even a new operating system entirely rather than junk your computer and start again with a new one when you need to make a change. As long as you understand how an operating system works and know how to get at it, you can in many cases change some of the ways it behaves.

The purpose of an operating system is to organize and control hardware and software so that the device it lives in behaves in a flexible but predictable way.

Not all computers have operating systems. The computer that controls the microwave oven in your kitchen, for example, doesn't need an operating system. It has one set of tasks to perform, very straightforward input to expect (a numbered keypad and a few pre-set buttons) and simple, never-changing hardware to control.

All desktop computers have operating systems. The most common are the Windows family of operating systems developed by Microsoft, the Macintosh operating systems developed by Apple and the UNIX family of operating systems. There are hundreds of other operating systems available for special-purpose applications, including specializations for mainframes, robotics, manufacturing, real-time control systems and so on.

Technical Words and Phrases

alphanumeric	[ælfənjuː'merɪk]	*adj.* 字母与数字混合编排的
assembly	[ə'semblɪ]	*n.* 集合，装配，汇编
bugs	[bʌgz]	*n.* 原因不明的故障　*adj.* 疯狂的，发疯的
compartmentalization	[ˌkɔmpɑːtˌməntlaɪ'zeʃən]	*vt.* 把……分成各自独立的部分，把……分成区；把……分门别类
debug	[dɪ'bʌg]	*vt.* 检测、跟踪并排除计算机程序或其他软件中的错误
discrete	[dɪ'skriːt]	*adj.* 离散的；分立的；不连续的
documentation	[dɔkjumən'teɪʃən]	*n.* 文件管理，文件编制
excessive	[ɪk'sesɪv]	*adj.* 过度的；极端的
hardware	['hɑːdweə(r)]	*n.* （计算机的）硬件，五金器件
impose	[ɪm'pəuz]	*vt.* 征税，强加，以……欺骗　*vi.* 利用，欺骗，实施影响
instruction	[ɪn'strʌkʃən]	*n.* 指令
microprocessor	[maɪkrəu'prəusesə(r)]	*n.* 微处理器
mnemonic	[nɪ'mɔnɪk]	*adj.* 记忆的，记忆术的　*n.* 记忆码
operand	[ɔpə'rænd]	*n.* 操作数，运算数，运算对象
plot	[plɔt]	*vt.* 绘制，标出
software	['sɔftweə(r)]	*n.* （计算机的）软件
subroutine	['sʌbruːˌtiːn]	*n.* 子程序
temporary	['tempərərɪ]	*adj.* 暂时的
variable	['veərɪəb(ə)l]	*n.* 变量　*adj.* 易变的；不稳定的，变量的

assembly mnemonic	汇编助记忆码
make sense	有意义
microwave oven	微波炉

operating system	操作系统
opcode	*n.* 操作代码，操作符，operating code 的缩写
source code	源代码

Notes to the Text

1. Algol 算法语言，一种以算法表示计算机程序的高级语言。其初始版本开发于 20 世纪 60 年代。后来的版本称为 Algol 68，主要用于数学和科学方面的应用领域。

2. Pascal 在计算机程序设计中，一种高度结构化的高级程序设计语言，是由瑞士苏黎世的 ETH 大学的 Niklaus Wirth 研制的。

3. block-structured language 模块结构化语言

10.2 Reading Materials

10.2.1 What Does Operating System Do

At the simplest level, an operating system does two things:

Firstly, it manages the hardware and software resources of the system. In a desktop computer, these resources include such things as the processor, memory, disk space, etc. On a cell phone, they include the keypad, the screen, the address book, the phone dialer, the battery and the network connection. Secondly, it provides a stable, consistent way for applications to deal with the hardware without having to know all the details of the hardware.

The first task, managing the hardware and software resources, is very important, as various programs and input methods compete for the attention of the central processing unit (CPU) and demand memory, storage and input/output (I/O) bandwidth for their own purposes. In this capacity, the operating system plays the role of the good parent, making sure that each application gets the necessary resources while playing nicely with all the other applications, as well as husbanding the limited capacity of the system to the greatest good of all the users and applications.

The second task, providing a consistent application interface, is especially important if there is to be more than one of a particular type of computer using the operating system, or if the hardware making up the computer is ever open to change. A consistent application program interface (API) allows a software developer to write an application on one computer and have a high level of confidence that it will run on another computer of the same type, even if the amount of memory or the quantity of storage is different on the two machines.

Even if a particular computer is unique, an operating system can ensure that applications continue to run when hardware upgrades and updates occur. This is because the operating system is charged with managing the hardware and the distribution of its resources. One of the challenges facing developers is keeping their operating systems flexible enough to run hardware from the thousands of vendors manufacturing computer equipment. Today's systems can accommodate thousands of different printers, disk drives and special peripherals in any possible combination.

10.2.2 Introduction of Microsoft Certification Program (MCP)

For all the tasks that are being performed by computer, someone needs to write the software to handle the task. This is the role of the software developer. Finding qualified software developers is a major issue for many companies today.

But how is a company supposed to determine who is qualified and who is just a computer user with a few elementary programming skills? Determining the qualifications of people is difficult when there are so many who claim to be computer experts, even if their expertise were only setting up their own web page. To help with the task of determining who is qualified and who is not, Microsoft developed its Microsoft certification program to certify those people who have the skills to work with Microsoft products and networks.

For developers, this certification is in the form of the Microsoft Certified Solution Developer. This certification establishes that a person is familiar with the inner workings of Windows and has skills in working with multiple programming languages or development environments.

The MCP program offers multiple certifications, based on different areas of technical expertise. To attain these certifications, a candidate must pass a series of exams within the program. The current certifications are Microsoft Certified Technology Specialist (MCTS), Microsoft Certified Professional Developer (MCPD), Microsoft Certified IT Professional (MCITP) and the Microsoft Certified Architect (MCA).

Popular previous generation certifications include Microsoft Certified System Engineer (MCSE), Microsoft Certified Solution Developer (MCSD) and Microsoft Certified Database Administrator (MCDBA)

10.2.3 Object-Oriented Programming

Ever since computers were used to solve application problems, it was the task of software to bridge the gap between the concepts in an application and computer concepts. The typical programming task is to translate the application concepts into computer concepts. If the translation succeeds and the computer solves the application problem, a successful software product has been developed.

Advances in software technology are driven by the desire to make the translation from application concepts to computer concepts easier. Advances in software technology have narrowed the gap from both sides. In the late 1960s it was recognized that a combination of both forces software design technology and advances in programming language might be possible. The resulting language construct was a class. A class implements the concepts of abstraction, modularization, and data hiding. It does so by grouping a user-defined type with all procedures and functions that can be applied to it. Classes allow inheritance. The concept of inheritance relates to the design concept of stepwise refinement and also allows the reuse of existing code and data structures in a class.

The engineering of a class concept represents a step to software technology similar to the development of an integrated circuit (IC) to hardware. A new hardware system can be built by using (or modifying) off-the-shelf ICs; it is now possible to build software systems by reusing (or extending) off-the-shelf classes, "Class" is the central concept in object-oriented programming.

The task at hand now is to educate the many programmers who still use yesterday's software technology to solve today's software problems. Object-oriented programming requires a new attitude toward problem solving. Problem solving with a computer system becomes more natural. Application concept can directly be translated into classes. It is commonly believed that it will be easier to train new programmers to use object-oriented techniques than to retrain programmers who have substantial experiences in applying conventional programming language constructs.

10.3 Knowledge about Translation
（翻译知识10——连词Ⅱ）

1．unless, once

unless 可译成除非、如果不，可引导条件状语从句；once 意为一旦、一经、只要……便、在……之后，作为连词可引导条件状语从句，或时间条件状语从句。

2．provided (that) 或 providing (that)

连词 provided (that) 意为"如果……、只要……，"可引导条件状语从句。

All substances will permit the passage of some electric current, **provided** the potential difference is high enough. 如果有足够高的电位差，一切物体都可以传导一些电流。

3．whether

whether 作为连词可引导主语从句，表语从句，宾语从句，同位语从句和让步状语从句。尤其是用做宾语从句和让步状语从句用得比较多，译成"是否……"。

The substances are divided into two classes, according to **whether** they did or did not electrify by rubbing. 物质可以根据能否摩擦起电而分成两大类。

whether 还可以引导宾语从句

Whether or not the speed is constant, the average speed is the distance the body moves divided by the time required for the motion. 不论速度是否恒定，平均速度都等于物体移动的距离除以运动所需的时间。

其中 the body moves 是定语从句，说明 distance，required for the motion 也是定语从句，说明 time。

4．although (though)

连词 although (though)的意思是"虽然"，"尽管"，可引导让步状语从句，although 让步语气较重，一般用于正式文章中，though 的让步语气较弱，常用于口语。

People sometimes confuse static electricity with magnetism, **though** the two are different. 尽管静电和磁是两种不同的现象，但有时人们却把它们两者混为一谈。

5．no matter how (what, where, who, which, when)

其意思是无论怎样（什么，什么地方，谁，哪一个，什么时候）引导让步状语从句。

It is through the electric circuit that energy is transmitted electrically from the primary source, **no matter where** it is situated to the ultimate consumer. 正是通过电路才能把电能从电源（无论

它位在何处）输送到最远的用户。

6. whenever, wherever, whatever, however

whenever 意思是"无论何时"，有时也可译成"一……就……"，"每当……"，"只要……"。

Whenever there is a current in a resistor, there is a drop in potential. 只要电阻中有电流，就会有电位降。这里 whenever 译成"只要……"，"每当……"，同样 wherever, whatever, however 可分别译成"无论何处，无论什么，无论怎样"。

10.4　Exercises

1. Put the Phrases into English

① 与……有些类似　　　　　　⑥ 副作用
② 以学术观点来看　　　　　　⑦ 易读的程序
③ 结构化语言　　　　　　　　⑧ 计算机高级语言
④ 显著特征　　　　　　　　　⑨ 汇编语言指令
⑤ 局部变量　　　　　　　　　⑩ 描述语句的注释

2. Put the Phrases into Chinese

① perform a specific task
② to creep into a program
③ in various situations
④ section off and hide from the rest of the program
⑤ provide an alternative to assembly language programming
⑥ correspond to one assembly language statement
⑦ using alphanumeric symbols
⑧ make it easier for programmers to write
⑨ two statement with the braces
⑩ create program structure

3. Translation

① The distinguishing feature of block-structured language is that the compartmentalization of code and data.
② Functions are the building blocks of C, in which all program activity occurs.
③ All variables that are declared in that function will be known only to that function.
④ Assembly language instructions are provided to describe each of the basic operation that can be performed by a microprocessor.
⑤ You will even find computers being used in most restaurants to handle order processing, a task that was previously done by hand.
⑥ The microprocessor is the hardware component. To get its work done, the microprocessor executes a set of instructions known as software.

⑦ Microsoft offers several levels of certification for anyone who has or is pursuing a career as a professional developer working with Microsoft products.

⑧ The most common are the Windows family of operating systems developed by Microsoft.

⑨ Ever since computers were used to solve application problems, it was the task of software to bridge the gap between the concepts in an application and computer concepts.

⑩ The task at hand now is to educate the many programmers who still use yesterday's software technology to solve today's software problems.

4. Read Object-Oriented Programming (10.2.3) and Choose the Best Answer

① What is the software's task for the computer to be used to solve an application problem?

   A. to bridge the gap between the concepts in an application and computer concepts.

   B. to direct the computer to treat the required application problem.

   C. to tell the computer exactly what to do.

   D. to guide the computer step by step to analyze the problem to be solved.

② We can say that a successful software product has been developed when

   A. the software developed has directed the computer to solve the problem successfully.

   B. the software developed has directed the computer to treat the required application problem successfully.

   C. the software developed has worked out the translation problem successfully.

   D. the translation from the application concepts into computer concepts of the problem to be solved is made successfully and the computer solves the application problem.

③ What is the desire for driving the advances in software technology?

   A. to narrow the gap between the application concepts and the computer concepts.

   B. to make easier the translation from application concepts to computer concepts.

   C. to provide software design principles and techniques that allow the programmer to apply.

   D. to allow the programmer to conceptualize application concepts and to simplify their transformation into computer system concepts.

④ Inheritance allows the programmers to

   A. use the existing code and data structure in an ancestor function

   B. refine the program in a class step by step

   C. refine the programs step by step and reuse the existing code and data structures in a class

   D. reuse the experience of other programmers

⑤ The engineering of a class concept represents

   A. a new design concept with which a new hardware system can be built by using (or modifying) off-the shelf ICs

   B. a new design concept with which a new software system can be built by reusing (or extending) off-the shelf ICs.

   C. a step to software technology like the development of an integrated circuit (IC) to hardware

   D. the central concept in object-oriented programming

10.5 课文参考译文

10.5.1 结构化语言C

与 Algol 和 Pascal 语言相似，C 通常被看成是一种结构化语言。虽然从学术观点来看，模块-结构化语言这个词并不严格适用于 C，C 只是模块-结构化语言中的一个非正式的部分。模块-结构化语言的主要特征是代码与数据的分离化。这意味着这种语言可以把完成一个指定任务所需要的信息和指令与程序的其他部分分离开并隐藏起来。一般说来，可以通过带有局部变量或临时变量的子程序实现程序与数据的分离，用这种方法，你可以编写一段子程序，所有在子程序中出现的操作将不会对程序的其他部分产生副作用。作为对照，全局变量是指在整个程序中都可以用的变量，过多地应用全局变量可能会在整个程序中引起错误和意想不到的副作用。在 C 语言中，所有的子程序都是独立的函数。

函数是 C 语言的组成模块，在函数中可以进行各种编程。函数可以使编程员在一个程序中分别定义和编码完成指定的任务。在调试过一个只用局部变量的函数以后，这个函数可以在各种情况下正常运用，不会对程序的其他部分产生副作用。所有在这个函数中声明的变量只能在这个函数中应用。

在 C 语言中，用代码块还可以创建程序结构。一个代码块是一组逻辑上相联系的程序语句，它们可以被看成是一个块。你可以在左右两个花括号之间插入几行代码创建一个代码块。如

```
if(x<10) {
    printf ("too low, try again");
    reset_counter(-1);
}
```

如果 x 小于 10，if 后面的位于花括号之间的两个语句就被执行，这两个句子连同两个花括号的语句表示一个代码块，这两个语句是连在一起的，不可能只执行其中一个语句。在 C 语言中，每个语句可以只有一个语句，也可以是一个语句块。利用代码块可以创建合乎逻辑的容易理解的程序。

C 语言是一种编程语言。不像其他高级计算机语言，C 语言对程序员能用它做什么没有什么限制。在绝大多数情况下，编程者可以使用 C 语言来替代汇编语言。事实上，发明 C 语言的原因之一就是为了提供一种替代汇编语言编程的方法。

10.5.2 MATLAB 语言

MATLAB 是一种数字化计算环境和可编程的语言。是数学工作室（MathWorks）开发的。MATLAB 集成了计算、可视化和编程在一个极易使用的环境中，问题和解答都以人们熟悉的数学符号表示。

MATLAB 是一个交互系统，其基本数据元素是阵列，用它可以很方便地计算很多问题，尤其是含矩阵和矢量公式的问题。另外它还有附加的 Simulink 软件包，增加了图形多领域仿真，以及 Model-Based Design 软件，用于动态和嵌入式系统。

MATLAB 用起来很方便，这里举两个例子。

1. 启动和退出

当启动 MATLAB 时，桌面上显示用于管理文件、变量以及和 MATLAB 相关的应用的工

具窗口（图形化用户界面）。

图 10.1 给出常用的窗口，你可以根据你的需要定制工具和文档。

要退出 MATLAB，在菜单中选择"文件">"退出 MATLAB"，或在命令窗口中直接输入：quit。

2. 画图

MATLAB 提供了很多方法显示图像数据，还有交互工具可以用来处理图像，使图像可以反映出关于数据的最多信息。

例如，下面的语句建立了一个变量 x（阵列），其值从-1 到 1，间隔为 0.1，第二句语句求出 x（阵列）的每一个值的三次方，并存储为 y 阵列：

  x = -1:.1:1; %定义 x 阵列
  y = x.^3; %求出 x（阵列）的每一个值的三次方，并存储为 y 阵列
  plot(x,y); %画 y-x 图像

一个简单的曲线（图）适当地显示了 x 作为自变量，y 作为函数的关系（如图 10.2 所示）。你还可以加上注释和打印此图，或输出图片的标准格式用以在网页浏览器或其他媒体上显示此图。

10.5.3 汇编语言指令

汇编语言指令是（厂家）提供的用来描述微处理器可以执行的最基本的指令。汇编语言通常用字母符号来代替微处理器的机器码。汇编语言语句通常放在左边，语句的注释则放在右边。这种格式使编程者容易书写、阅读和调试代码。词"代码"是指用微处理器语言书写的程序（即机器码）。用汇编语言书写的程序被称做源代码。

源程序中的每一条指令都与一条汇编语言语句一一对应。语句必须指明执行什么操作和对什么数据进行操作。因此，一条指令可以分成两个部分：操作码和操作数。操作码是指出要执行的操作，如典型的操作是加、减和移动。

在汇编语言中，我们指定一个、两个或三个字母的组合来表示一种操作，这些字母组合是为了记住指令。如在 8086 汇编语言中加、减和移动的指令分别是 ADD, SUB 和 MOV。

操作数指出用微处理器执行操作码指定的操作时要处理的数据，如在把 B 寄存器中的数与 A 寄存器中的数相加的指令中，BX 和 AX 是操作数。描述这条指令的汇编语言是：

  ADD AX, BX

80x86 指令（大致）可以分为八类：

① 数据移动指令 MOV, LEA, LES , PUSH, POP, PUSHF, POPF
② 转换指令 CBW, CWD, XLAT
③ 算术指令 ADD, INC SUB, DEC, CMP, NEG, MUL, IMUL, DIV, IDIV
④ 逻辑、移位、转动和位操作指令 AND, OR, XOR, NOT, SHL, SHR, RCL, RCR
⑤ 输入/输出指令 IN, OUT
⑥ 字符串指令 MOVS, STOS, LODS
⑦ 程序控制指令 JMP, CALL, RET, CONDITIONAL JUMPS
⑧ 其他指令 CLC, STC, CMC

10.5.4 操作系统简介

当前台式计算机和小型计算机通常有一个微处理器作为它的中央处理器，微处理器是硬

件，工作时，微处理器执行一系列的指令，这些指令称为软件。有两种不同的软件：

操作系统——操作系统为计算机中运行的应用程序提供一系列的服务，还为计算机提供基本的用户接口。

应用程序——应用程序是为实现某个具体任务而编写的程序软件，现在你的计算机中可能就有浏览应用程序、字处理应用程序、电子邮箱应用程序等，你也可以购买安装新的应用程序。

近年来，操作系统在小型设备中开始流行。现在这些小型设备中所用的微处理器功能已很强了，所以也可以运行操作系统和应用软件了。现在一般手机中的微处理器比20年前的台式计算机功能还要强大，所以（使用操作系统）是有意义的，也是一个很自然的发展。在任何一个有操作系统的电气设备中，通常都可以设置这些电气设备的工作方式。采用操作系统并不是偶然的，是因为操作系统是通过代码而不是通过（不变）硬件电路来设置电器（工作方式）的，所以调整（电器工作方式）时不用拆开整台电器。

对一个台式机用户来说，这就意味着你可以设新的密码、加系统补丁、安装新的应用程序甚至新的操作系统，当你想对这些设置进行改变时，完全不需要换一台新的计算机。只要你了解操作系统是如何工作的和如何使用操作系统，你就可以调整操作系统的部分性能。

操作系统的目的是组织和控制硬件和软件，使设备的工作方式是可调节和预先设定的。

并不是所有的计算机都有操作系统，例如厨房中控制微波炉的微处理器就可能不需要操作系统，它只有一组可执行的任务，只要直接输入数字键和一些预设置的按钮后，就可以让硬件电路去控制（微波炉）完成指定的任务了。

所有的台式机都有操作系统，最通用的是微软开发的 Windows 系列的操作系统、苹果公司开发的 Macintosh 操作系统和 UNIX 系列操作系统。还有数百种其他的操作系统是为包括大型机、机器人、制造系统和实时控制系统等专门应用目的而设计的。

10.6 阅读材料参考译文

10.6.1 操作系统做些什么

一个操作系统至少要做这样两件事：

一是管理系统的硬件和软件资源。在台式机中，这些资源包括处理器、内存、硬盘等，在手机中，这些资源包括键盘、显示屏、地址本、电话号码、电池和网络连接。二是操作系统为应用程序提供了一种稳定、一致的接口（界面），使得处理硬件的应用程序并不需要知道具体的硬件电路。

第一个任务即管理硬件和软件资源是非常重要的，因为各种程序和输入方法为各自的目的争抢 CPU、内存、存储设备和输入/输出口带宽，这时操作系统起到一个好的管理作用，它保证各个应用程序得到必要的资源并与其他应用程序协调工作，操作系统面向所有的用户和应用程序，使系统的有限资源得到最充分的应用。

第二个任务即提供一个一致的应用界面，尤其对不同类型的计算机或硬件不断变化的计算机来说采用操作系统就显得更加重要。一个一致的应用程序界面（API）使一个软件开发者在一台计算机上编写的应用程序可以在其他任何同类的计算机上运行，即使两台计算机的内存或存储量不同也没关系。

即使对同一台计算机，操作系统也保证它在硬件升级和更新后应用程序仍可以运行。因为操作系统是负责管理硬件和资源分布的。开发者要解决的是保持他们的操作系统可适用于来自于几千个计算机设备制造商的硬件。现在的操作系统可适用于几千个不同的打印机、硬盘驱动器和专用外围设备的各种任意组合。

10.6.2 微软认证项目介绍

目前所有用计算机处理的工作都需要人们编写（计算机）软件来处理，这就是软件开发人员的工作。今天对于许多公司来说找到称职的软件开发人员是很重要的一件事。

但是一个公司如何来确定某人是很称职的，而某人只是一个具有少量的基本编程技能的计算机用户呢？当有很多人甚至其专长只会设置他们的个人网页就声称自己是计算机专家时，确定人们的称职与否是很难的。为此，微软设立了它的微软认证项目，以确认那些具有微软产品和网络开发技能的人。

对开发者来说，这种认证是以获得"微软认证软件开发专家"证书的形式被认可的，这个证书确保（具有此证书的人）熟悉 Windows 的内部工作原理，有能力运用多种编程语言编程或开发环境。

微软认证项目根据不同领域对技术专业知识的要求提供多种证书，为了获得这种证书，必须通过一系列的项目考试。目前的证书有微软认证技术专家（MCTS）、微软认证专业开发人员（MCPD）、微软认证 IT 专业人员（MCITP）和微软认证设计师（MCA）。

以前很流行的证书包括微软认证系统工程师（MCSE）、微软认证软件开发专家（MCSD）和微软认证数据库管理专家（MCDBA）。

10.6.3 面向对象的编程

自从计算机用来解决各种实际应用问题至今，软件的任务就是在实际应用的概念和计算机概念之间架起一座桥梁。典型的编程任务就是把实际应用概念转化成计算机概念。如果转化得很成功，则计算机解决了实际应用问题，一个成功的软件产品就开发出来了。

要把从实际概念转化成计算机概念的过程变得更加容易的愿望促进了软件技术的进步。软件技术的进步使实际概念与计算机概念的差距变小了。在 20 世纪 60 年代后期，人们意识到软件设计技术和编程语言的改进是可以相结合的，导出的语言结构就是"类"。"类"通过组合（封装）用户定义的（数据）类型以及处理这些数据的所有程序和函数实现了抽象、模块化和数据隐藏。类有继承性，继承的概念与逐步求精的设计概念有关，继承也使得一个类中已有的代码和数据结构可得到再利用。

与硬件中集成电路的发展一样，一个"类"概念的形成表明了软件技术的一大进步。一个新的硬件系统可以通过用（或修改）现成的集成电路来建立，同样现在新的软件系统也可以通过重新使用或扩展现成的类来构造了，"类"是面向对象程序设计的核心思想。

当前的任务是要教育那些为数众多的至今还在用昨天的软件技术解决今天的软件问题的程序员。面向对象的程序设计技术要求采取新的态度去解决问题。用计算机系统解决问题变得更简单，应用的概念可以直接转变成类。通常认为训练一个新的程序员去使用面向对象技术比重新训练那些对使用传统程序设计语言结构有丰富经验的程序员更为容易。

Unit 11 Television

> *Pre-reading*
>
> Read the following passage, paying attention to the question.
> 1) What is CRT?
> 2) What difference between the color TV and black-white TV?
> 3) How do you get the signal to TV?

11.1 Text

11.1.1 about Television

Television is certainly one of the most influential forces of our time.

To understand TV, let's start at the beginning with a quick note about your brain. There are two amazing things about your brain that make television possible. The first principle is this: If you divide a still image into a collection of small colored dots, your brain will reassemble the dots into a meaningful image. On a TV or computer screen, the dots are called pixels.

The human brain's second amazing feature relating to television is this: If you divide a moving scene into a sequence of still pictures and show the still images in rapid succession, the brain will reassemble the still images into a single, moving scene.

1. the Cathode Ray Tube

Some TVs in use today rely on a device known as the cathode ray tube, or CRT, to display their images. Let's start with the CRT, because CRTs are used to be the most common way of displaying images.

In a cathode ray tube (see in Fig 11.1), the "cathode" is a heated filament (not unlike the filament in a normal light bulb). The heated filament is in a vacuum created inside a glass "tube." The "ray" is a stream of electrons that naturally pour off a heated cathode into the vacuum.

Electrons are negative. The anode is positive, so it attracts the electrons pouring off the cathode. In a TV's cathode ray tube, the stream of electrons is focused by a focusing anode into a tight beam and then accelerated by an accelerating anode. This tight, high-speed beam of electrons flies through the vacuum in the tube and hits the flat screen at the other end of the tube. This screen is coated with phosphor, which glows when struck by the beam.

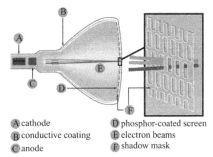

Ⓐ cathode   Ⓓ phosphor-coated screen
Ⓑ conductive coating Ⓔ electron beams
Ⓒ anode   Ⓕ shadow mask

Fig 11.1   CRT

There are coils, which are able to create magnetic fields inside the tube. One set of coils creates a magnetic field that moves the electron beam vertically, while another set moves the beam horizontally. By controlling the voltages in the coils, you can position the electron beam at any point on the screen.

In a CRT, phosphor coats the inside of the screen. When the electron beam strikes the phosphor, it makes the screen glow. In a black-and-white screen, there is one phosphor that glows white when struck. In a color screen, there are three phosphors arranged as dots or stripes that emit red, green and blue light. There are also three electron beams to illuminate the three different colors together.

2. the Black-and-White TV Signal

In a black-and-white TV, the electron beam "paints" an image onto the screen by moving the electron beam across the phosphor a line at a time. As the beam paints each line from left to right, the intensity of the beam is changed to create different shades of black, gray and white across the screen. Because the lines are spaced very closely together, your brain integrates them into a single image. A TV screen normally has about 480 lines visible from top to bottom.

When a television station wants to broadcast a signal to your TV, or when your VCR(Video Cassette Recorder) wants to display the movie on a video tape on your TV, the signal needs to mesh with the electronics controlling the beam so that the TV can accurately paint the picture that the TV station or VCR sends. The TV station or VCR therefore sends a well-known signal to the TV that contains three different parts:

- Intensity information for the beam as it paints each line
- Horizontal-retrace signals to tell the TV when to move the beam back at the end of each line
- Vertical-retrace signals 60 times per second to move the beam from bottom-right to top-left

11.1.2 Color TV

A color TV screen differs from a black-and-white screen in three ways:

• There are three electron beams that move simultaneously across the screen. They are named the red, green and blue beams.

• The screen is not coated with a single sheet of phosphor as in a black-and-white TV. Instead, the screen is coated with red, green and blue phosphors arranged in dots or stripes. If you turn on your TV or computer monitor and look closely at the screen with a magnifying glass, you will be able to see the dots or stripes.

• On the inside of the tube, very close to the phosphor coating, there is a thin metal screen called a shadow mask. This mask is perforated with very small holes that are aligned with the phosphor dots (or stripes) on the screen (Fig 11.2).

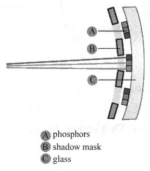

Ⓐ phosphors
Ⓑ shadow mask
Ⓒ glass

Fig 11.2   how the shadow mask works

When a color TV needs to create a red dot, it fires the red beam at the red phosphor. Similarly for green and blue dots. To create a white dot, red, green and blue beams are fired simultaneously - the three colors mix together to create white. To create a black dot, all three beams are turned off as they scan past the dot. All other colors on a TV screen are combinations of red, green and blue.

A color TV signal starts off looking just like a black-and-white signal. An extra chrominance signal is added by superimposing a 3.579 545 MHz sine wave onto the standard black-and-white signal. A phase shift in the chrominance signal indicates the color to display. The amplitude of the signal determines the saturation. The Tab 11.1 shows you the relationship between color and phase.

Tab 11.1 a phase shift in the chrominance

Color	Phase
Burst	0 degrees
Yellow	15 degrees
Red	75 degrees
Magenta	135 degrees
Blue	195 degrees
Cyan	255 degrees
Green	315 degrees

A black-and-white TV filters out and ignores the chrominance signal. A color TV picks it out of the signal and decodes it, along with the normal intensity signal, to determine how to modulate the three color beams.

11.1.3 Getting the Signal to TV

You are probably familiar with five different ways to get a signal into your TV set:

① Broadcast programming received through an antenna.

The composite TV signal can be broadcast to your house on any available channel. The composite video signal is amplitude-modulated into the appropriate frequency, and then the sound is frequency-modulated as a separate signal.

② VCR or DVD player that connects to the antenna terminals.

VCRs are essentially their own little TV stations. The video tape contains a composite video signal and a separate sound signal. The VCR has a circuit inside that takes the video and sound signals off the tape and turns them into a signal that, to the TV, looks just like the broadcast signal for channel 3 or 4.

③ Cable TV arriving in a set-top box that connects to the antenna terminals.

The cable in cable TV contains a large number of channels that are transmitted on the cable. Your cable provider could simply modulate the different cable TV programs onto all of the normal frequencies and transmit that to your house via the cable; then, the tuner in your TV would accept the signal and you would not need a cable box. Unfortunately, that approach would make theft of cable services very easy, so the signals are encoded in funny ways. The set-top box is a decoder. You select the channel on it, it decodes the right signal and then does the same thing a VCR does to transmit the signal to the TV.

④ Large (6 to 12 feet) satellite-dish antenna arriving in a set-top box that connects to the antenna terminals.

Fig 11.3   small-dish satellite

Large-dish satellite antennas pick off unencoded or encoded signals being beamed to Earth by satellites. First, you point the dish to a particular satellite, and then you select a particular channel it is transmitting. The set-top box receives the signal, decodes it if necessary and then sends it to TV.

⑤ Small (1 to 2 feet) satellite-dish antenna arriving in a set-top box that connects to the antenna terminals.

Small-dish satellite systems (Fig 11.3) are digital. The TV programs are encoded in MPEG-2 format and transmitted to Earth. The set-top box does a lot of work to decode MPEG-2, then converts it to a standard analog TV signal and sends it to your TV.

Technical Words and Phrases

antenna	[æn'tenə]	n. 天线
cable	['keɪb(ə)l]	n. 电缆，海底电报，此处指有线电视
channel	['tʃænəl]	n. 海峡，信道，频道　vt. 引导，开导，形成河道
chrominance	['krəumɪnəns]	n. 色度（任意一种颜色与亮度相同的一个指定的参考色之间的差异，如彩色电视采用白色为参考色）
dot	[dɔt]	n. 点，圆点　vt. 在……上打点
filament	['fɪləmənt]	n. 灯丝；细丝
intensity	[ɪn'tensɪtɪ]	n. 强烈，剧烈，强度，亮度
phosphor	[fɔsfə(r)]	n. 磷，启明星
pixel	['pɪks(ə)l]	n.（显示器或电视机图像的）像素
plasma	[plæzmə]	n. 等离子体，等离子区，这里指等离子显示器
satellite	['sætəlaɪt]	n. 人造卫星
saturation	['sætʃə'reɪʃ(ə)n]	n. 饱和（状态），浸润，浸透，饱和度
sequence	['siːkwəns]	n. 次序，顺序，序列
succession	[sək'seʃ(ə)n]	n. 连续，继承，继任，演替
vacuum	['vækjuəm]	n. 真空，真空吸尘器　adj. 真空的　vt. 用真空吸尘器打扫

covert…into　　　　　　　　　把……转换
integrate…into…　　　　　　　把……整体组合成……
VCR(Video Cassette Recorder)　盒式录像机，磁带式录像机

11.2 Reading Materials

11.2.1 Digital TV

The horizontal resolution is something like 500 dots for a color analog TV set. This level of resolution was amazing 50 years ago, but today it is rather passed. The lowest resolution computer monitor that anyone uses today has 640×480 pixels, and most people use a resolution like 800×600 or 1024×768. We have grown comfortable with the great clarity and solidity of a computer display, and analog TV technology pales by comparison.

Many of the new satellite systems, as well as DVDs, use a digital encoding scheme that provides a much clearer picture. In these systems, the digital information is converted to the analog format to display it on your analog TV. The image looks great compared to a VHS tape, but it would be twice as good if the conversion to analog didn't happen.

There is now a big push underway to convert all TV sets from analog to digital, so that digital signals drive your TV set directly.

When you read and hear people talking about digital television (DTV), what they are talking about is the transmission of pure digital television signals, along with the reception and display of those signals on a digital TV set. The digital signals might be broadcast over the air or transmitted by a cable or satellite system to your home. In your home, a decoder receives the signal and uses it, in digital form, to directly drive your digital TV set.

There is a class of digital television that is getting a lot of press right now. It is called high-definition television, or HDTV. HDTV is high-resolution digital television combined with Dolby Digital surround sound (AC-3). This combination creates a stunning image with stunning sound. HDTV requires new production and transmission equipment at the HDTV stations, as well as new equipment for reception by the consumer. The higher resolution picture is the main selling point for HDTV. Imagine 720 or 1080 lines of resolution compared to the 525 lines people are used to in the United States (or the 625 lines in Europe) - it's a huge difference!

11.2.2 LCD (Liquid Crystal Display)

You probably use items containing an LCD (liquid crystal display) every day. They are all around us - in laptop computers, digital clocks and watches, microwave ovens, CD players and many other electronic devices. LCDs are common because they offer some real advantages over other display technologies. They are thinner and lighter and draw much less power than cathode ray tubes (CRTs).

One feature of liquid crystals is that they're affected by electric current. A particular sort of nematic liquid crystal, called twisted nematics (TN), is naturally twisted. Applying an electric current to these liquid crystals will untwist them to varying degrees, depending on the current's voltage.

There's far more to building an LCD than simply creating a sheet of liquid crystals. The combination of four facts makes LCD possible:

- Light can be polarized.
- Liquid crystals can transmit and change polarized light.
- The structure of liquid crystals can be changed by electric current.
- There are transparent substances that can conduct electricity.

An LCD is a device that uses these four facts in a surprising way!

There are two main types of LCDs used in computers, passive matrix and active matrix.

Most LCD displays use active matrix technology. A thin film transistor (TFT) arranges tiny transistors and capacitors in a matrix on the glass of the display. To address a particular pixel, the proper row is switched on, and then a charge is sent down the correct column. Since all of the other rows that the column intersects are turned off, only the capacitor at the designated pixel receives a charge. The capacitor is able to hold the charge until the next refresh cycle.

The other type of LCD technology is passive matrix. This type of LCD display uses a grid of conductive metal to charge each pixel. Although they are less expensive to produce, passive matrix monitors are rarely used today due to the technology's slow response time and imprecise voltage control compared to active matrix technology.

11.2.3 Pure Vision Plasma Display

We'll just say this up front: you may have difficulty believing that your new PureVision Plasma Display (Fig 11.4) is only 4" thick and hanging on your living room wall, that you can watch DVD's on it, plus high-definition TV, home movies, and videotapes, and that you can play video games on it, surf the web, and do e-mail, but it's all true.

Fig 11.4 Plasma TV

One thing that you will believe immediately is how incredible the picture is; high-definition broadcasts and DVD movies look absolutely stunning, and when you're playing an analog source like a videotape or regular TV, the Pure Vision Plasma Display up-converts these signals to digital, for a vastly improved picture. When you're watching a movie, the display's PureCinema II™ function detects the film-based original source and instantly recreates each individual still frame, for a more film-like presentation.

When you're using it as a monitor for a Mac or PC, the PureVision Plasma Display's resolution is just as astounding, able to process full-specification VGA, SVGA, and XGA input.

The unit's back panel provides the breadth of connections you'll need for your various components, including those for a PC or Mac, and component and composite video inputs for a high-definition tuner, DVD player, or other sources.

注：Mac 和 PC 都是指个人计算机，Mac 是苹果公司的个人计算机。

（本文摘自于网站上的一则广告）

11.3 Knowledge about Translation
（翻译知识 11——虚拟语气）

虚拟语气是当作者想表示主观愿望和假想虚拟的情况或表示一种委婉的口气时，用谓语动词来表达的一种形式。在科技英语中也常有这种情况，翻译中注意要与实际的情形加以区分。

1. 虚拟条件从句

虚拟条件从句是指假设的情况根本不可能存在，或者发生的可能性很小。这时采用虚拟语气。它的基本形式有如下两种：

① 表示与现在的事实相反的假设：

从句	主句
谓语动词用过去时，be 一律用 were	Should (would, could, might) + 动词原形

If no force acted on the moon it would move in a straight line with constant speed.
如果月球不受力的作用，它就会以恒速作直线运动。
这里的条件是根本不可能存在的。

② 表示与过去的事实相反的假设：

从句	主句
Had + 过去分词	Should (would,could,might) +have+ 动词过去分词

If there had been no radio and television, man could not have seen the far side of the moon.
如果没有无线电和电视，人们就不能看到月球的背面了。

③ 省略 if 的倒装虚拟条件句：

在 if 引导的条件状语从句中有 were, had 时，有时可把 if 省略，而采用倒装语序，此时把 were, had 提到主语之前。

Had radio not been invented, many jobs, such as helping aeroplanes to land in bad weather and calling doctors to far-off places, would not have been possible. 如果没有发明无线电，很多工作如在恶劣的天气下帮助飞机着陆，请医生到遥远的地方去等就做不成了。

④ 引导虚拟条件从句的连词：

通常用 if 作为虚拟条件从句的连词，此外 suppose, provided (that), once, unless 等连词都可引导虚拟条件从句。

From Lenz's Law we knew that neither electrical nor other form of energy can be derived unless an equivalent amount of work be performed. 根据楞次定律我们知道，如果不做等量的功，就既不能得到电能也不能得到其他形式的能量。

unless 引导的条件从句用的是虚拟语气，谓语是 be performed。

Normally, DC is not useful on a transformer unless it be varying.一般直流电在一个变压器中是没有用的，除非直流电是在变化的。

2．谓语动词一律用原形的虚拟条件句

在表示现在或将来有可能实现的一种假想时，从句中一切人称都用动词原形，而主句中的谓语则用陈述语气，这种用法在科技英语中也很常见。

It follows that those points must be at the same electrical level if there be no movement of electricity between any two points. 如果任意两点间没有电荷流动，则这两点必然处于相同的电位。

3．虚拟假设情况的其他表示法

虚拟条件句的假设情况还可用下列含有条件意味的词、短语或句子来表示：

or less（否则，要不然），without（如果没有），otherwise（否则），in the obsence of （如果不存在，如果没有）等等。

How would you get money out of the bank on a Sunday night without electronics? 如果没有电子技术你怎么可能在星期日的晚上从银行（自动取款机）中取出钱？

Without such an increase in productivity that the design of complex systems might not be achievable within a reasonable time frame. 如果没有生产率的提高，在合理的时间框架内（或译成：在近期内）复杂系统的设计就不能实现。

4. 虚拟语气用于一些从句的情况

① 用于表示愿望、建议、要求等动词后的宾语从句常用虚拟语气：

英语中有些及物动词，如 wish, desire, require, demand, think, expect，suggest, propose, imagine 等，具有愿望、建议、命令的含义，它们后面的从句中谓语动词常用(should)+动词原形。且一般都省略 should。

② as if, as though（好像，仿佛）引导的方式方法状语从句用虚拟语气。

You can be heard just as clearly as if the two of you were in the same room.
声音听起来很清楚，就好像你们两人在同一个房间里一样。

11.4 Exercises

1. Translate the Following Phrases into English

① 静止图像　　　　　　⑥ 建立电磁场
② 与电视有关的　　　　⑦ 水平移动光束
③ 最常见的显示图像方式　⑧ 电子束
④ 阴极射线管，显像管　⑨ 发射红、绿、蓝光
⑤ 像素　　　　　　　　⑩ 电信号

2. Translate the Following Phrases into Chinese

① reassemble the dots into a meaningful image
② second amazing feature
③ hit the flat screen
④ contain three different parts
⑤ the three colors mix together
⑥ satellite broadcast
⑦ the normal intensity signal
⑧ be familiar with…
⑨ cable TV programs
⑩ select the channel

3. Sentence Translation

① To understand TV, let's start at the beginning with a quick note about your brain.

② LCDs and plasma displays are sometimes seen, but they are still rare when compared to CRTs.

③ There are coils, which are able to create magnetic fields inside the tube.

④ When the electron beam strikes the phosphor, it makes the screen glow.

⑤ As the beam paints each line from left to right, the intensity of the beam is changed to create different shades of black, gray and white across the screen.

⑥ A color TV screen differs from a black-and-white screen in three ways.

⑦ When a color TV needs to create a red dot, it fires the red beam at the red phosphor, similarly for green and blue dots.

⑧ You are probably familiar with five different ways to get a signal into your TV set.

⑨ The composite video signal is amplitude-modulated into the appropriate frequency, and then the sound is frequency-modulated as a separate signal.

⑩ Unfortunately, that approach would make theft of cable services very easy, so the signals are encoded in funny ways.

4. Translation

① Clearly, Philips is ready for the future.

When friends drop by, they are amazed at my new Philips Digital TV. The Real Flat picture tube and amazing images give a film-like viewing experience that is second to none.

② Sometimes a TV just isn't a TV.

Not my Philips TV. Dual Tuner PIP(双重调谐的尖峰信号), Digital Comb Filter for excellent pictures, Component Video for optimum picture, Incredible Surround Sound, remote to control all my accessories. No. It's not just a TV. It's everything I could have wanted and more.

（两条从飞利浦网站上摘录的数字电视机广告语）

11.5　课文参考译文

11.5.1　关于电视

电视肯定是我们这个时代最有影响力的东西之一。

要了解电视，先看一下人的大脑。大脑中有两样奇妙的东西使电视成为可能，第一就是如果你把一幅静止的图片切割成很多小彩色点，你的大脑会把这些彩色的点拼成一幅（有意义的）图像。在电视机和计算机显示屏上，这些点被称做像素。

人脑的第二个与电视有关的奇妙之处在于如果你把一幅运动的图像分解成一连串静止画面并以很快且连续的方法显示这些静止的画面，大脑会把这些静止的画面重新组合成一个动态的场面。

1. 阴极射线管

现在所用的电视机有一部分是用阴极射线管（CRT）来显示图像的，我们先讨论 CRT，因为 CRT 曾经是最普通的显示器。

在一个阴极射线管中（如图 11.1 所示），"阴极"是加热的灯丝（与普通白炽灯的灯丝不同）。加热灯丝放在一个玻璃真空管中，"射线"是一束电子束，是从加热灯丝射入真空管的。

电子是负电荷，阳极是正的，所以阳极吸引电子离开阴极，在一个电视机的阴极射线管中，电子流被聚焦阳极聚焦成一细（电子）束，加速阳极使电子束加速。这一高速电子束穿过真空撞击在真空管另一端的平面屏幕上，这个屏幕上镀了一层磷，当电子束撞击它时会发光。

在真空管中有可以产生电磁场的线圈，一组线圈产生使电子作垂直运动的电磁场，另一组线圈产生使电子作水平运动的电磁场。通过控制线圈的电压，就可以使电子束撞击到屏幕的任意位置。

在 CRT 中，屏幕的内表面镀了一层磷，当电子撞击在磷上，会使屏幕发光。黑白电视机屏幕只有一层撞击时发白光的磷，彩色电视机屏幕有排列成点状或条状的三种磷，分别发红光、绿光和蓝光，还需要三个电子光束同时使这三种不同颜色发光。

2. 黑白电视信号

在黑白电视机中，电子束通过每次沿着一条磷线移动电子束在屏幕上"画"出一幅图像。当电子束每次从左边扫到右边时，电子束的强度是变化的，在屏幕上显示不同的黑、灰、白点。因为这些线排列很密，你的大脑把它们组合成一幅图像。一般一个电视机屏幕从顶部到底部大约有 480 条线。

当用电视机播放电视台传送的信号或录像机中录像带上的电影时，信号要用电路划分成可以控制光束的（像点）信号，则电视机就可以正确地显示电视台或录像机送来的图像，电视台或录像机发送给电视机的信号中含有以下三个部分：

- 光束扫描的强度。
- 水平返回信号即控制电视机的电子束扫描到每行的终端并返回的信号。
- 垂直返回信号——每秒 60 次把电子束从右下端移到左上端。

11.5.2 彩色电视

一个彩色电视屏幕与黑白电视屏幕有三点不同：

- 有三个电子束同时在屏幕上扫描，分别为红色、绿色、蓝色电子束。
- 屏幕上不是像黑白电视机那样只镀一层磷，而是镀了排列成点状或条状的红色、绿色和蓝色的磷。如果打开电视机或计算机显示器用放大镜凑得很近去看屏幕，可以看到这些点或条。
- 在阴极射线管的内部，在磷点附近，有一细金属屏，称为荫罩板，荫罩板上有很多细孔与屏幕上的磷光点（或条）同样排列（如图 11.2 所示）。

当彩色电视机要产生一个红点时，它射出红色电子束撞击红光磷，产生绿点和蓝点也是同样的原理。要产生一个白点，则红、绿、蓝三个电子束同时射出，三种颜色混合在一起产生白点。要产生一个黑点，所有三色电子束都在扫过这点时关掉，电视机上其他颜色是红、绿、蓝的组合。

表 11.1 部分颜色和相位的关系

颜色	相位
白	0 度
黄	15 度
红	75 度
洋红（红紫色）	135 度
蓝	195 度
青（蓝绿色）	255 度
绿	315 度

初看起来彩色电视信号与黑白电视信号有些相似，但要在标准的黑白电视信号上多叠加一个 3.579 545 MHz 的正弦波传送一个色度信号。色度信号中相位的不同表示出要显示的颜色，色度信号的幅度决定了颜色的饱和度。表 11.1 表示颜色和相位的关系。

黑白电视机会滤去这些色度信号，彩色电视机则取出这些信号并解调，配合一般的强度信号，去调制三个彩色电子束（显示彩色图像）。

11.5.3 电视机接收到的信号

你可能对电视机可接收到的五种信号很熟悉：

① 通过天线接收到的电视台无线信号。

复合的电视信号可通过适当的无线频道传送到家中，混合的电视信号是调幅的视频信号和独立的调频声音信号。

② 录像机或 DVD 播放机与天线接口相连接。

录像机本身基本上是一个小电视台，录像带包含了合成视频信号和独立的音频信号，录像机内部有一个电路可把磁带上的视频信号和音频信号取出并转换成一个类似于电视台（3或4频道）发送出的电视信号，送入电视机。

③ 有线电视机顶盒与（电视机的）天线接口相连接，接收有线电视信号。

有线电视电缆中含有大量正在传输的频道，有线电视台可简单地把不同的有线电视台的节目调制成各种频率信号通过电缆传送到你家。然后电视机的调谐器将接收信号，这样并不需要机顶盒。但这样很容易就可以偷取有线电视台的服务（即不付钱收看），所以电视信号以一种特殊（古怪）的方式编码，机顶盒再解码。你在机顶盒中选某个台，它就把这个台的信号解码，再像录像机一样把信号发送到电视机中。

④ 大卫星天线（6～12英尺）传输到机顶盒，再接到天线接口。

大卫星天线接收来自卫星定向传到地球的编码或不编码的信号，首先，用天线对准某一卫星，选择这卫星所传输信号的频道，机顶盒接收到信号，若需要解码先进行解码再送入电视机。

⑤ 小卫星天线（1～2英尺）传输到机顶盒，再接到天线接口。

小卫星系统（如图11.3所示）是数字式的，电视节目用MPEG-2（一种压缩格式）格式编码，传送到地球上，机顶盒把MPEG-2格式的信号解压解码，再转换成标准的模拟信号送入电视机。

11.6　阅读材料参考译文

11.6.1　数字电视

一个模拟彩电的水平分辨率约500点，这个分辨率在50年前是很高的，但今天却过时了。现在所用的计算机显示器最低分辨率为640×480像素，很多人用的是800×600或1024×768像素的显示器，计算机的显示器正在变得更加清晰和可靠，相比之下，模拟电视机就比较逊色了。

许多新的卫星（电视）系统以及DVD，是采用可以提供十分清晰图像的数字编码机制，这种系统中，数字信息被转换成模拟格式，在模拟电视机上显示。与VHS式图像相比，这种图像很清晰，但如果不把这些图像转换成模拟格式，图像将会更加清晰。

现在，正开始逐渐把所有模拟电视机转换成数字电视机，这样就可以直接输出数字信号。

当提到数字电视时，指的是纯数字电视信号传输以及用数字电视机接收并播放这些数字信号。数字信号可以无线传播，也可以通过有线及卫星传播，在家中通过一个解调器以数字格式接收数字信号并直接在数字电视机上播放。

现在有一种很吸引人的数字电视，称为高清电视（HDTV），它是高分辨率数字电视与Dolby格式环绕数字立体声（AC-3）的结合，这个结合创造了最好的图像和最好的声音。HDTV要求HDTV电视台有新的制作和传输设备，还要求用户有新的接收设备。高分辨率图像是HDTV主要卖点，720或1 080线分辨率的图像与目前美国525线分辨率（或欧洲625线分辨率）的图像相比，图像绝对要清晰得多。

11.6.2　LCD（液晶显示器）

在台式计算机、数字钟表、微波炉、CD播放器和许多其他日用电子产品中都含有液晶显示器（LCD）。LCD之所以如此普及是因为与其他显示技术相比它确实有优点。它比CRT

薄、轻且功耗小。

液晶的一个特点是液晶容易受电流影响，一种特殊的向列液晶，称扭曲向列，本身是扭曲的，在这种液晶中有电流流过时，液晶根据所加的电压改变其扭曲程度。

并不是简单地铺一层液晶就可以做成 LCD，做成一个 LCD 要结合以下 4 点：
- 偏振光的产生。
- 液晶可以传输和改变偏振光。
- 液晶的结构可以用电流来改变。
- 有可以导电的透明物质。

LCD 就是奇妙地结合了这 4 种技术的产品。

计算机所用的 LCD 主要有两种类型：无源矩阵和有源矩阵。

大部分 LCD 显示器用有源矩阵技术，在显示器的玻璃上铺了一层以矩阵形式排列的小晶体管和电容构成晶体管薄膜（TFT），为定位某个像素，相应行的开关合上，然后把电荷送到相应的列上。因为该列上其他点的行开关都断开，只有指定像素点的电容接收电荷。电容可以保持这个电荷直到下一次刷新。

另一种 LCD 技术是无源矩阵技术，这种 LCD 显示器用一个导电金属栅给各像素点充电，虽然制造成本比较低，但现在无源矩阵很少用，因为与有源矩阵技术相比这种技术时间响应慢，且电压控制不够精确。

11.6.3 纯平、等离子显示器

我们先声明：你可能很难想象新的纯平、等离子显示器（如图 11.4 所示）只有 4 英寸厚，可以挂在你的客厅的墙上，通过它你可以看 DVD 和高清晰度的电视节目、家庭影院和录像带，你也可以用它玩视频游戏、上网浏览、发电子邮件，但这些都是真的。

但有一点你可以马上看到该显示器上的图像是多么令人不可思议的清晰，高清晰度的节目和 DVD 电影看起来绝对刺激（或译为绝对使人震惊），当你播放一个模拟（信号）源的节目如录像带或一般的电视节目时，纯平、等离子显示器把这些（模拟）信号转换成数字信号，并更好地改善了图像的质量。当你看电影时，显示的 PureCinema II™（电影纯平化）功能探测到胶片上的原信号并立即再创建出每一个独立的静止帧，以供更多的胶片一样地放映。

当你用它作为个人计算机的显示器时，纯平等离子显示器的分辨率是令人吃惊的，它可以处理所有规格的 VGA, SVGA 和 XGA 输入。

后板上提供了各种各样的接口，可用来接各种部件，包括与个人计算机连接的输入接口、作为高清晰度调谐的合成视频（信号）输入接口、DVD 播放器接口或其他输入源接口。

Unit 12 Digital Camera

> ***Pre-reading***
>
> *Read the following passage, paying attention to the question.*
> *1) What is digital photography?*
> *2) What does the picture quality of a digital camera depend on?*
> *3) Could you tell some features about the digital camera?*
> *4) What is CF card?*

12.1 Text

12.1.1 Principle

In principle, a digital camera is similar to a traditional film-based camera. There's a viewfinder to aim it, a lens to focus the image onto a light-sensitive device, some means by which several images can be stored and removed for later use, and the whole lot is fitted into a box. In a conventional camera, light-sensitive film captures images and is used to store them after chemical development. Digital photography uses a combination of advanced image sensor technology and memory storage, which allows images to be captured in a digital format that is available instantly - with no need for a "development" process.

Although the principle may be the same as a film camera, the inner workings of a digital camera are quite different (Fig 12.1). The imaging being performed either by a charge coupled device (CCD) or CMOS (complementary metal-oxide semiconductor) sensors. Each sensor element converts light into a voltage proportional to the brightness, which is passed into an analogue-to-digital converter (ADC), which translates the fluctuations of the CCD into discrete binary code. The digital output of the ADC is sent to a digital signal processor (DSP), which adjusts contrast and detail, and compresses the image before sending it to the storage medium. The CCD or CMOS sensors are fixed in place and it can go on taking photos for the lifetime of the camera. There's no need to wind film between two spools either, which helps minimize the number of moving parts.

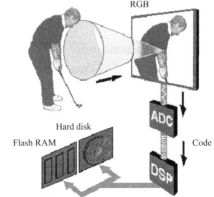

Fig 12.1 the workings in a digital camera

12.1.2 Picture Quality

The picture quality of a digital camera depends on several factors, including the optical quality of the lens and image-capture chip, compression algorithms, and other components. However, the most important determinant of image quality is the resolution of the CCD. The more elements, the higher the resolution, and thus the greater the detail that can be captured. The first consumer model 4 megapixel camera appeared in mid-2001, boasting a maximum image size of 2240×1680 pixels.

The quality of a CCD's colour management process is another important factor and one of the prime reasons for differences in the output of cameras with the same pixel count CCD. The process should not be confused with the interpolation method used by some manufacturers to achieve bitmap files with a resolution greater than their true optical resolution (the resolution of their CCD array).

Another limiting factor is the image compression routines used by many digital cameras to enable more images to be stored in a given amount of memory. Most digital cameras compress and save their images in the industry-standard JPEG or FlashPIX formats, readable on almost every graphics package. Both use slightly lossy compression leading to some loss of image quality. However, many cameras have several different compression settings, allowing the user a trade-off between resolution quality and image capacity, including the option to store images in with no compression at all ("CCD raw mode") for the very best quality.

12.1.3 Features

A color LCD panel is a feature that is present on virtually all modern digital cameras. It acts as a mini GUI, allowing the user to adjust the full range of settings offered by the camera and is an invaluable aid to previewing and arranging photos without needing to connect to a PC to do so. Typically this can be used to display some thumbnails of the stored images simultaneously, or provide the option to view a particular image full-screen, zoom in close and, if required, delete it from memory.

Digital cameras offer two distinct varieties of zoom feature: optical zoom and digital zoom. Optical zoom works in much the same way as a zoom lens on a traditional camera. Produced by the lens system, it is the magnification difference between minimum and maximum focal lengths. Importantly, in digital cameras this magnification occurs before an image is recorded in pixels. Digital zoom, on the other hand, is arguably little more than a marketing gimmick. Digital zoom is nothing more than the cropping of the middle of an image by a digital camera's software. When an image that has been digitally zoomed 2× is reproduced, either on a display monitor or by being printed, it will effectively be viewed at half its original resolution.

Most digital cameras offer a number of image exposure timing options. One of the most popular is a burst mode that allows a number of exposures to be taken with a single press of the shutter. The speed and number of sequential shots that can be captured in a burst is dependent on the amount of internal memory the camera possesses, the image size selected and the degree of compression applied to the photos.

Features allowing a variety image effects are becoming increasingly common. For example, a user may have the option to select between monochrome, negative and sepia modes. Apart from their use for artistic effect, the monochrome mode is useful for capturing images of documents. Some digital cameras also provide a "sports" mode - which adds sharpness to the captured images of moving objects - and a "night shooting" mode, which allows for long exposures.

A self-timer is a common feature, typically providing a 10-second delay between the time the shutter is activated and when the picture is taken and all modern day digital cameras have a built-in automatic flash, with a manual override option. The best have a working range of up to 12ft and provide a number of different modes.

12.1.4 Memory and Connectivity

Many first-generation digital cameras contained one or two megabytes of internal memory suitable for storing around 30 standard-quality images at a size of 640×480 pixels.

By early 1999 two rival formats were battling for domination of the digital camera arena.

CompactFlash (CF): CF provides non-volatile storage that doesn't require a battery to retain data. It's essentially a PC flash card that's been reduced to about one quarter of its original size and uses a 50-pin connection that fits into a standard 68-pin Type II PC Card adapter. By late 2008 maximum capacities had reached 16 GB.

SmartMedia (SM): The Toshiba-developed SmartMedia cards are significantly smaller and lighter than CompactFlash cards, Capacities are less than for CompactFlash −128 MB was still the maximum capacity by late 2001, capable of storing 560 high-resolution (1 200×1 024) still photographs - and cost per megabyte is similar to that of CompactFlash.

In August 1999, Panasonic, SanDisk, and Toshiba first agreed to develop and market the Secure Digital Memory Card (SD) (Fig 12.2), Today SD card is widely used in digital cameras, handheld computers, PDAs, mobile phones, GPS receivers, and video game consoles. Standard SD card capacities range from 4 MB to 4 GB, and for high capacity SDHC cards from 4 GB to 32 GB as of 2008.

Fig 12.2 SDcard

Devices with SD slots can use the thinner MMCs, but standard SD cards will not fit into the thinner MMC slots. miniSD and microSD cards can be used directly in SD slots with a simple passive adapter, since the cards differ in size and shape but not electrical interface. With an active electronic adapter, SD cards can be used in CompactFlash or PC card slots.

Despite the trend towards removable storage, digital cameras still allow connection to a PC for the purpose of image downloading. Until the late 1990s the principal method of transfer was via a conventional RS232 serial cable at a maximum speed of 115 Kb/s. However, since then USB connectivity has become the norm, with most manufacturers bundling cameras with the necessary cables and driver software.

USB is also exploited by a method of image transfer that emerged in the early 2000s and that saves the bother of having to connect a camera to a PC at all. These "media readers" are available for all the common media types - CompactFlash, SD Card etc. - and simply plug in to a USB port, either directly or via an extension cable.

Technical Words and Phrases

compress	[kəm'pres]	vt. 压缩，摘要叙述
consumer	[kən'sjuːmə(r)]	n. 消费者；用户
contrast	[kən'trɑːstj]	vt. 使与……对比，使与……对照　vi. 和……形成对照，n. 对比度
distinct	[dɪ'stɪŋkt]	adj. 清楚的，明显的，截然不同的，独特的
exploit	[ɪk'splɔɪt]	vt. 开拓，开发，开采，剥削，用以自肥 v. 使用
exposure	[ɪk'spəʊʒə(r)]	n. 暴露，揭露，曝光，揭发
film	[fɪlm]	n. 薄膜，胶卷，影片，电影
lens	[lenz]	n. 透镜；镜头，（眼睛的）晶体
lossy	[lɔsɪ]	adj. 有损耗的，致损耗的
magnification	[mægnɪfɪ'keɪʃ(ə)n]	n. 扩大；放大，放大率，放大倍数
manufacturer	[mænju'fæktʃərə(r)]	n. 制造者，制造商，制造
medium	[miːdɪəm]	n. 媒体，方法，媒介　adj. 中间的，中等的，半生熟的
minimize	['mɪnɪmaɪz]	vt. 将……减到最少　v. 最小化
monitor	['mɒnɪtə(r)]	n. 班长，监听器，监视器　vt. 监控　v. 监视
monochrome	['mɒnəkrəʊm]	n. 单色画；单色照片；黑白相片
negative	['negətɪv]	n. 否定，负数，底片　adj. 否定的，阴性的　vt. 否定
panel	['pæn(ə)l]	n. 面板，嵌板，仪表板　vt. 嵌镶板
photography	[fə'tɒgrəfɪ]	n. 摄影，摄影术
prime	[praɪm]	n. 最初，青春，精华　adj. 主要的，最初的
raw	[rɔː]	adj. 未加工的，生疏的，处于自然状态
resolution	[rezə'luːʃ(ə)n]	n. 分辨率，决定，分解，坚定，决心，决议
sensitive	['sensɪtɪv]	adj. 敏感的，灵敏的，感光的
sepia	['siːpɪə]	n. 棕褐色
storage	['stɔːrɪdʒ]	n. 存储，储藏（量），储藏库
thumbnail	['θʌmneɪl]	n. 拇指甲，极小的东西　adj. 极小的，极短的
viewfinder	['vjuːfaɪndə(r)]	n.（照相机）取景器
zoom	[zuːm]	n. 急速上升，图像电子放大，缩放，变焦　vi. 突然扩大，摄像机移动　vt. 使摄像机移动

analogue-to-digital converter (ADC)	模-数转换器
charge coupled device (CCD)	电荷耦合器件
compression algorithms	（图像）压缩格式
digital signal processor (DSP)	数字信号处理器
industry-standard JPEG format	工业标准格式
RS232 serial port	RS232（协议）串行口

12.2　Reading Materials

12.2.1　Digital Camcorders

As recently as the first half of the 1990s few would have dreamed that before long camcorders would be viewed as a PC peripheral and that video editing would have become one of the fastest growing PC applications. All that changed with the introduction of Sony's DV format in 1995 and the subsequent almost universal adoption of the IEEE 1394 interface, making a digital camcorder almost as easy to attach to a PC system as a mouse.

Suddenly enthusiasts had access to a technology that allowed them to produce source material in a digital format whose quality far exceeded that of the analogue consumer formats available at the time - such as Hi-8 and S-VHS - and to turn this into professional-looking home movies at their desktop PC.

DV cassettes won't play in VCRs, of course, but any digital camcorder will include conventional, analogue AV output jacks to allow recorded material to be transferred to VCR or viewed on a TV set. Many mainstream consumer digital camcorders are sold as all-in-one solutions for video, stills and even MP3 and email. Most can only capture stills at a resolution similar to that of DV video - 720×576 pixels - a resolution that is usually reduced to 640×480 in order to retain the correct aspect ratio. Some camcorders boast higher resolutions for stills, but often these larger images have been interpolated to reach the specified resolution.

12.2.2　Video Compression

Video compression is the art of throwing as much data away as possible without it showing. The technology by which video compression is achieved is known as a "codec", an abbreviation of compression/decompression. Various types of codec have been developed - implemental in either software and hardware, and sometimes utilizing both - allowing video to be readily translated to and from its compressed state.

Lossy techniques reduce data - both through complex mathematical encryption and through selective intentional shedding of visual information that our eyes and brain usually ignore - and can lead to perceptible loss of picture quality. "Lossless" compression, by contrast, discards only redundant information. Codecs have compression ratios ranging from a gentle 2:1 to an aggressive 100:1, making it feasible to deal with huge amounts of video data. The higher the compression ratio, the worse the resulting image.

By the end of the 1990s, the dominant techniques were based on a three-stage algorithm known as DCT (Discrete Cosine Transform). DCT uses the fact that adjacent pixels in a picture - either physically close in the image (spatial) or in successive images (temporal) - may be the same value. A mathematical transform - a relative of the Fourier transform - is performed (Fig 12.3) on grids of 8×8 pixels (hence the blocks of visual artefacts at high compression levels). It doesn't reduce data but the resulting coefficient frequency values are no longer equal in their information-carrying roles. Specifically, it's been shown that for visual systems, the lower frequency components are more important than high frequency ones. A quantisation process weights these accordingly and ejects those contributing least visual information, depending on the compression level required. For instance, losing 50 per cent of the transformed data may only result in a loss of five percent of the visual information. Then entropy encoding - a lossless technique - jettisons any truly unnecessary bits.

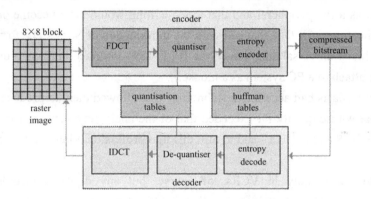

Fig 12.3　compression/decompression card

12.2.3　X3 Technology

In 2002 the prospect of truly affordable film-quality digital cameras was given a massive boost when - after five years of research and development - Foveon Corporation unveiled a digital camera imaging sensor which the company claimed was capable of obviating 35mm film.

In conventional digital cameras systems colour filters are applied to a single layer of photo-detectors in a tilted mosaic pattern. The filters let only one wavelength of light - red, green or blue (Fig 12.4) - pass through to any given pixel, allowing it record only one colour. As a result, typical mosaic sensors capture 50% of the green and only 25% of each of the blue and red light. The approach has inherent drawbacks, no matter how many pixels a mosaic-based image sensor might contain. Since they only capture one third of the colour, mosaic-based image sensors must rely on complex processing to interpolate the two-thirds they miss. Not only does this slow down the speed of image rendering, interpolation also leads to colour artefacts and a loss of image detail. Some cameras even intentionally blur pictures to reduce colour artefacts.

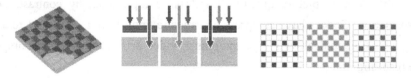

Fig 12.4　the single layer of photo-detectors

Foveon's new CMOS image sensor uses the company's revolutionary X3 technology to capture up to three times more information per pixel than modern-day digital cameras at similar megapixel resolutions. The X3 image sensors accomplish this by using three layers of photo detectors embedded in silicon (Fig 12.5). The layers are positioned to take advantage of the fact that silicon absorbs different colors of light at different depths, so one layer records red, another layer records green and the remaining layer records blue. This means that for every pixel on a Foveon X3 image sensor, there's actually a stack of three photo detectors. The result is a sensor capable of capturing red, green, and blue in each pixel location - in essence, the first full-colour digital camera image sensor.

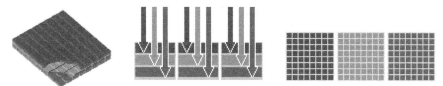

Fig 12.5　three layers of photo detectors in an X3 image sensor

12.3 Knowledge about Translation
（翻译知识12——倒装）

英语句子与中文一样，通常是主语在前，谓语其次，宾语在后，这种语序称为自然语序，否则称为倒装语序。使用倒装语序通常是为了强调句中的某个成分，或是由于句子结构的要求。如英语的问句一般为倒装句，英语一些习惯用法也用倒装句。一般的倒装句如问句，there be 句型这里不再讨论，这里讨论科技英语中常用且在翻译时容易出错的倒装句型。

1. 句首有表示否定的词

在英语中，当一些含有否定意义的词放在句首时，一般要用倒装结构。这些词有：not, no, never, hardly（几乎不）, seldom（很少）, scarcely（几乎不）, not until, not only, neither, nor , in no way, no longer（不再）, nowhere 等。

Not only **can** electricity be made to produce magnetism, but magnetism can be made to produce electricity. 不但电可以产生磁，磁也可以产生电。

2. here ,only, so, then 等词放在句首

按英语习惯用法，把 here, only, so 等词放在句首时，句子采用倒装结构。

Steel is a good conductor of electricity, and so **are** most metals. 钢是良导体，大多数金属也是良导体。

Only when steel is heated to the critical temperature **does** the grain become very fine. 只有把钢加热到临界温度时，晶粒才变得很细。

Here is the bus. 公共汽车来了。

3. 某些状语从句

如 no matter how, no matter what, however, whatever 等连词引导的让步状语从句，常把所修饰的宾语放在最前面。

Electrons are universal constituents of matter, their mass and electric charge being the same from whatever **element** they are obtained. 电子是物质的普遍成分，无论哪种元素中电子的质量和电荷都是相同的。

As (though)引导的让步状语从句结构则为：表语＋as(though)＋主语＋助动词，如

Light as aluminum is , its strength is great. 铝虽然轻，但强度很高。

当 as 引导的从句中有助动词时，可以把动词放在前面。

Try as you may, you cannot be successful. 你可以去试试，但你不会成功的。

有时甚至让步状语从句不用连词，而用倒装句，这时谓语的一部分通常是 be，且用原形，放在句首，翻译时要注意。

Nearly all our clothes are made from fibre of one sort or another, **be** they derived from where. 我们所有的衣服几乎都是用某种纤维制成的，无论这些纤维来源于何处。

as 或 just as 引导的方式状语从句和 than 引导的比较状语从句一般用倒装句，句中的谓语往往用 do 代替主句中的实义动词。

We may conclude that air occupies space, just as **does** any other fluid. 我们可以得出结论，空气像任何流体一样也占据空间。

The molecules of the liquid have more energy than **do** the molecules of the solid. 液体中的分子比固体中的分子具有更多的位能。

12.4 Exercises

1. Translate the following phrases into English

① 数码相机
② 光敏器件
③ 先进的图像传感器技术
④ 压缩图像
⑤ 最大图片尺寸
⑥ 数码相机的分辨率
⑦ 另一个重要的因数
⑧ 光学变焦
⑨ 有损压缩
⑩ 存储图像

2. Translate the following phrases into Chinese

① traditional film-based camera
② with no need for a "development" process
③ convert light into a voltage
④ adjust contrast and detail
⑤ wind film between two spools either
⑥ image-capture chip
⑦ some loss of image quality
⑧ view a particular image full-screen
⑨ digital zoom
⑩ memory card

3. Sentence Translation

① Although the principle may be the same as a film camera, the inner workings of a digital camera are quite different.

② The picture quality of a digital camera depends on several factors, including the optical quality of the lens and image-capture chip, compression algorithms, and other components.

③ The more elements, the higher the resolution, and thus the greater the detail that can be captured.

④ Another limiting factor is the image compression routines used by many digital cameras to enable more images to be stored in a given amount of memory.

⑤ For example, a user may have the option to select between monochrome, negative and sepia modes.

⑥ Many first-generation digital cameras contained one or two megabytes of internal memory suitable for storing around 30 standard-quality images at a size of 640×480 pixels.

⑦ Despite the trend towards removable storage, digital cameras still allow connection to a PC for the purpose of image downloading.

⑧ A self-timer is a common feature; typically providing a 10-second delay between the time the shutter is activated.

⑨ Video compression is the art of throwing as much data away as possible without it showing.

⑩ For instance, losing 50 per cent of the transformed data may only result in a loss of five per cent of the visual information.

4. Translation

In recent years, the Application Specific Integrated Circuits (ASIC) market has been dominated by gate arrays and their improved form-the sea of gates (SOG) arrays. The latter provides the advantages of quick turnaround times, high packing density and high performance circuits. With the introduction of large, channelless SOG arrays, conventional routers may no longer be able to handle the ever –increasing complexity of the VISI interconnection problem.

12.5　课文参考译文

12.5.1　原理

原理上，数字相机与传统的胶片相机（光学相机）相似：有一个可对准目标的视窗，一个光学镜头把图像聚焦在光敏器件上，保存图像的方法，以便以后取出图像使用，以及所有这些都封装在一个盒（照相机）中。（不同的是）在传统的照相机中是底片感光捕获图像，通过冲洗胶卷来保存这些图像。数码照相则是先进的图像传感器技术和存储技术的结合，它以数字格式捕获图像，这些图像无须"冲洗"就可以得到。

虽然原理与胶片照相机相似，但数码相机的内部工作情况完全不同（如图12.1所示），图像用一个电荷耦合器件（CCD）或一个CMOS传感器捕获，每点的传感器把光转换成与光亮度成正比的电压，再把这个电压送入模数转换器（ADC），而模数转换器把CCD送来的波动电压信号转换成离散的二进制代码。ADC输出的数字代码又送到数字信号处理器（DSP），DSP调整对比度和清晰度（细节），压缩数据，再把它送到存储介质中。CCD或CMOS传感器是固定的，（只要照相机不坏）它们可以继续拍摄，也不需要卷（两个轴之间的）胶卷，因此可以减少照相机中运动的部件。

12.5.2 图像质量

数码相机的照片质量由这样一些因素决定：光学镜头的质量、捕获图像的芯片、压缩方法和其他部件。但决定图像质量最重要的因数是 CCD 的分辨率。点数越多，分辨率越高，则更多的图像细节被捕获到（或译照片更清晰）。2001 年中期造出第一台 400 万像素的数码相机，厂商自吹最大可以拍摄 2240×1680 像素的照片。

CCD 的色彩处理过程的质量是另一个重要的因素，相同像素 CCD 的照相机的输出图像可能因此而不同。不要把这个过程与插值方法相混淆，插值方法是指一些厂商所采用的使照片的分辨率达到高于照片真正光学分辨率（CCD 分辨率）的一种（计算机）处理方法。

还有一个限制的因素是图像的压缩方法，许多数码相机为了能在给定的存储卡中存储更多的图像采用压缩方法。大部分数码相机以工业标准 JPEG 或 FlashPIX 格式压缩图像，几乎所有的图像软件都可以观看这些图像，但这两种格式都是有损压缩，会造成图像质量受损。但很多照相机有一些不同的压缩比设置，允许用户在分辨率质量和图像大小之间作一选择，包括选择用完全不压缩（CCD 原模式）格式存储图像，以得到质量非常好的照片。

12.5.3 特点

现在所有的数字相机都有一个彩色 LCD（液晶显示屏），这是一个小型图形界面，用户可以调节照相机的全部设置，有了它，用户不用将数码相机连接到计算机就可以预览和整理照片。典型地可选择同时观看数张小照片或选择观看一张照片全景或局部放大，需要的还可以从存储器中删除照片。

数字相机有两个变焦参数：光学变焦和数字变焦。光学变焦与传统照相机中的变焦镜头的工作方式很相像，由镜头系统组成，光学变焦是指最大焦距和最小焦距之间的放大倍数，更重要的这是数码相机在把图像记录成像素前的放大倍数。而数字变焦，则只是一个（销售）市场玩的花招。数字变焦就像用数字相机软件裁剪一张照片，当一张照片用数字变焦放大两倍，无论在显示屏上或印出的照片，其有效分辨率只是原来的一半。

许多数字照相机还有照片曝光时间选择，最常用的是一种连续曝光模式即只要按一下快门就可以连续拍摄若干张照片。连续拍摄的速度和数量由数码相机内部的存储量、选定的图像大小和对照片的压缩程度决定。

具有各种图像效果的特点现在也是日益普及，例如用户可以选择黑白、底片效果、棕褐色（或称老照片）模式，黑白模式除了有艺术效果外，还可以拍文档的照片。有些数码相机还有在拍运动物体的照片时增加锐度的"运动"模式和可以长时间曝光的"夜视"模式。

自拍也是数码相机都有的特点。一般是按下快门后延时 10 s 拍照。现在所有的数码相机都有内置闪光灯，有一个手动（闪光）模式选择钮。好的相机闪光范围达 12 英尺且有很多模式。

12.5.4 存储器和连接

很多第一代的数码相机中含有 1～2 MB 的内存，可存 30 张标准质量的 640×480 的照片。到 1999 年早期数码相机领域中主要是两种格式存储（卡）在竞争。

压缩闪存（CF）卡：CF 卡是一种不挥发的存储卡，不需要电池维持数据，它本来是为计算机设计的闪存卡，体积减小到原来的 1/4，用 50 脚的连接器固定到标准的 68 脚计算机

II 型适配器上。到 2008 年后期其最大容量达 16 GB。

SmartMedia（SM）卡：东芝开发的 SM 卡比 CF 卡小且轻，容量小于 CF 卡——2001 年后期最大容量仍只有 128 MB，可以存储 560 张高分辨率（1 200×1 024）静态照片，每兆的成本与 CF 卡差不多。

1999 年 8 月，松下、SanDisk 和东芝 3 家公司达成协议开发安全数字存储卡（SD 卡）（如图 12.2 所示），今天，SD 卡已广泛用于数码相机、手提电脑、电子记事本（个人计算机助理）、手机、GPS 接收器和视频游戏机中。标准的 SD 卡容量从 4 MB 到 4 GB。2008 年，已有容量可以从 4 GB 到 32 GB 的高容量的 SD 卡（写做 SDHC）。

很多带有 SD 插槽的设备可能用的是比较狭窄的 MMC 槽（SD 卡是在 MMC 的基础上开发的），因此标准的 SD 卡不能插入，但小型和微型的 SD 卡（SD 卡有 3 种型号）可以通过一个简单的无源适配器直接插到这些 SD 卡槽中，因为这些卡的尺寸和形状虽不同，但引脚定义相同。SD 卡还可以通过一个有源的适配器插在 CF 卡槽或 PC 卡槽中。

尽管目前的趋势是采用可移动存储器，但数码相机仍可与计算机连接直接下载照片。以前主要的传递方法是通过传统的 RS232 串行口，最大传输速率是 115 Kb/s。但到 1990 年后期，USB 接口成为标准接口，许多厂商制造的照相机都可以与 USB 接口连接并带有连接线和驱动软件。

在 21 世纪初期还可以把 USB 接口作为照片传输的一种方法，它免去了必须把照相机与计算机连在一起（才能把照片存储）的麻烦，"媒体读卡器"（有称"数码伴侣"）可以从任何普通存储类型（设备）——CF 卡、SD 卡等中读取图片，只要简单地把照相机直接或通过连接线插入 USB 口。

12.6　阅读材料参考译文

12.6.1　数码摄像机

在 20 世纪 90 年代中期以前几乎没人想到不久以后数码摄像机会成为计算机的外围设备，视频编辑会成为成长最快的计算机应用技术。但随着 1995 年索尼的 DV 格式的引入和紧接着的 IEEE1394 接口的普遍应用，使得数码相机就像一个鼠标一样很容易地与计算机相连接，所有这一切都发生了变化。

突然间，大家都热衷于这种技术，这种技术使他们可以用数字格式制作（视频）原材料，其质量远超过当时的模拟格式如 Hi-8,S-VHS 格式的图像，并用他们的台式计算机把它制作成准专业的家庭电影。

当然，DV 摄像带不能用录像机播放，但任何数码摄像机带有传统的模拟 AV 输出口端口，可以把录像内容传送给录像机或用电视机来看。许多主流数码摄像机产品都是集视频、静态（照片），甚至 MP3 和电子邮件于一体。大部分摄像机只能用接近它们的视频分辨率（720×576 像素）产生 640×480 的照片（为保持长、宽比）。有些摄像机厂商称对静态照片有更高的分辨率，但这些照片是通过插值的方法来达到所指的分辨率的。

12.6.2　视频压缩

视频压缩是指尽可能多地删去不影响显示的数据的技术，视频压缩的技术是通过压缩解压卡（简称 codec）实现的。目前已开发出各种压缩解压卡——软件解压卡、硬件解压卡和结合软硬技术的解压卡，使视频数据可以很方便地压缩或解压。

有损压缩技术是通过复杂的数学编码和故意有选择地去掉一些我们眼睛和大脑不注意的视频信息这两种方法压缩数据，可能会导致图像质量有所下降。作为对照，无损压缩只去掉冗余的信息。压缩解压卡有不同的压缩比，从较小的 2:1 压缩比到极大的 100:1，使得处理巨量的视频数据成为可能。压缩比越高，则有可能导致图片的质量也越差。

到 20 世纪 90 年代末期，主要的技术是基于三级运算的 DCT（离散余弦变换）压缩。DCT 利用图片中邻近的像素（不管是同一图像上邻近，还是连续图像之间的邻近）可能是相同的数据，对一个 8×8 像素的图片进行类似傅里叶变换的数学变换（如图 12.3 所示），（因此在高压缩比时就人为造成视频的误差），这一步并不减少数据，但导致在信息传输时（不同像素）的频率系数不再相等。可以证明，在视频系统中，低频分量比高频分量重要一些。在数据转换中比较这些数据的重要性，根据压缩比的要求删去最不重要的视频信息，例如删去 50%的输送数据可能只造成 5%的视频信息损失。然后对这些数据进行（平均）信息编码（一种无损技术），丢弃那些真正不需要的信息。

12.6.3　X3 技术

2002 年，经过 5 年的研究开发后，Foveon 公司推出一种声称可以替代 35 mm 胶卷的数码相机图片传感器。这时可拍出高质量（胶卷质量）照片的数码相机出现了，其前景广泛。

在传统的数码相机系统中，图像感光器的单层感光元素上覆盖一层马赛克状（镶嵌拼花式）的彩色滤波器，每个滤波器只允许一种波长的光（红、绿、蓝，如图 12.4 所示）通过到达指定的像素，像素只记录一种颜色，因此，传统的图像感光器接收到 50%绿光、25%蓝光和 25%红光。不管这种感光器有多少像素，它有天生的缺点，因为它只接收到颜色的 1/3，这种感光器要采用很复杂的插值处理方法来补充丢掉的另外 2/3 的数据。这不但降低了图像处理速度，而且还导致色彩失真和图片细节的损失，有些相机为减小色彩失真甚至故意使照片显得朦胧。

Foveon 的新的 CMOS 图像传感器用公司的 X3 技术在同样分辨率的前提下每个像素可以获取三倍于现数码相机（单层感光元素）的信息，嵌在硅中的 X3 图像传感器用三层感光元素来实现这一点（如图 12.5 所示）。按不同深度的硅吸收不同颜色的光来布置这三层，所以一层吸收红光，另一层吸收绿光，还有一层吸收蓝光。Foveon X3 感光器的每个像素都采用这种方法，实际上有三个感光器叠在一起。所以传感器每个像素位置都可以吸收红、绿、蓝三色，本质上，这是第一个全彩数码相机传感器。

Unit 13　Internet-based Communication

> **Pre-reading**
>
> *Read the following passage, paying attention to the question.*
> *1) What does the Internet offer to communicate?*
> *2) What is VoIP?*
> *3) Which defect in E-mail is referred to in the text?*

13.1　Text

If you use the Internet, then you probably use Internet-based communications to contact family, friends or co-workers. From sending an instant message to a friend, to E-mailing co-workers, to placing phone calls, to conducting videoconferences, the Internet offers a number of ways to communicate.

The advantages of Internet-based communications are many. Since you're already paying for an Internet account (or your employer is), you can save money on phone calls by sending someone an instant message or by using VoIP instead of standard local telephone services. Of course, no technology is without a downside and Internet-based communications has plenty, such as viruses, privacy issues and spam.

Like all technologies (and especially technology tied to the Internet), the way we can communicate online is constantly evolving. Here we'll take a look at some of the most popular forms of Internet-based communications.

13.1.1　Instant Messaging

Fig 13.1　IM-QQ

One of the fastest-growing forms of Internet communications is instant messaging, or IM (Fig 13.1). Think of IM as a text-based computer conference between two or more people. An IM communications service enables you to create a kind of private chat room with another individual in order to communicate in real-time over the Internet. Typically, the IM system alerts you whenever somebody on your buddy or contact list is online. You can then initiate a chat session with that particular individual.

While IM is used by millions of Internet users to contact family and friends, it's also growing in popularity in the business world. Employees of a company can have instant access to managers and co-workers in different offices and can eliminate the need to place phone calls when information is required immediately. Overall, IM can save time for

· 161 ·

employees and help decrease the amount of money a business spends on communications.

13.1.2 Internet Telephony & VoIP (Voice over Internet Protocol)

Internet telephony consists of a combination of hardware and software that enables you to use the Internet as the transmission medium for telephone calls. For users who have free, or fixed-price Internet access, Internet telephony software essentially provides free telephone calls anywhere in the world. In its simplest form, PC-to-PC Internet telephony can be as easy as hooking up a microphone to your computer and sending your voice through a cable modem to a person who has Internet telephony software that is compatible with yours. This basic form of Internet telephony is not without its problems. However, connecting this way is slower than using a traditional telephone, and the quality of the voice transmissions is also not near the quality you would get when placing a regular phone call.

Many Internet telephony applications are available. Some, such as CoolTalk and NetMeeting, come bundled with popular Web browsers. Others are stand-alone products. Internet telephony products are sometimes called IP telephony, Voice over the Internet (VoI) or Voice over IP (VoIP) products.

VoIP is another Internet-based communications method which is growing in popularity. VoIP hardware and software work together to use the Internet to transmit telephone calls by sending voice data in packets using IP rather than by traditional circuit transmissions, called PSTN (Public Switched Telephone Network). The voice traffic is converted into data packets then routed over the Internet, or any IP network, just as normal data packets would be transmitted. When the data packets reach their destination, they are converted back to voice data again for the recipient. Your telephone is connected to a VoIP phone adapter (considered the hardware aspect). This adapter is connected to your broadband Internet connection. The call is routed through the Internet to a regular phone jack, which is connected to the receiver's phone. Special hardware (the phone adapter) is required only for the sender.

Much like finding an Internet service provider (ISP) for your Internet connection, you will need to use a VoIP provider. Some service providers may offer plans that include free calls to other subscribers on their network and charge flat rates for other VoIP calls based on a fixed number of calling minutes. You most likely will pay additional fees when you call long distance using VoIP. While this sounds a lot like regular telephone service, it is less expensive than traditional voice communications, starting with the fact that you will no longer need to pay for extras on your monthly phone bill.

13.1.3 E-mail

Short for electronic mail, E-mail is the transmission of messages over communications networks. Most mainframes, minicomputers and computer networks have an E-mail system. Some E-mail systems are confined to a single computer system or network, but others have gateways to other computer systems, enabling you to send electronic mail anywhere in the world.

Using an E-mail client (software such as Microsoft Outlook or Eudora), you can compose an E-mail message and send it to another person anywhere, as long as you know the recipient E-mail address. All online services and Internet Service Providers (ISPs) offer E-mail, and support gateways so that you can exchange E-mail with users of other systems. Usually, it takes only a few seconds for an E-mail to arrive at its destination. This is a particularly effective way to communicate with a group because you can broadcast a message or document to everyone in the group at once.

One of the biggest black clouds hanging over E-mail is spam. Though definitions vary, spam can be considered any electronic junk mail (generally E-mail advertising for some product) that is sent out to thousands, if not millions, of people. Often spam perpetrates the spread of E-mail Trojans and viruses. For this reason, it's important to use an updated anti-virus program, which will scan your incoming and outgoing E-mail for viruses.

13.1.4 Videoconference

Videoconference is a conference between two or more participants at different sites by using computer networks to transmit audio and video data. Each participant has a video camera, microphone and speakers connected on his or her computer. As the two participants speak to one another, their voices are carried over the network and delivered to the other's speakers, and whatever images appear in front of the video camera appear in a window on the other participant's monitor.

In order for videoconference to work, the conference participants must use the same client or compatible software. Many freeware and shareware videoconference tools are available online for download, and most Web cameras also come bundled with videoconference software. Many newer videoconference packages can also be integrated with public IM clients for multipoint conferencing and collaboration.

In recent years, videoconference has become a popular form of distance communication in classrooms, allowing for a cost efficient way to provide distance learning, guest speakers, and multi-school collaboration projects. Many feel that videoconference provides a visual connection and interaction that cannot be achieved with standard IM or E-mail communications.

13.1.5 SMS & Wireless Communications

Short message service (SMS) is a global wireless service that enables the transmission of alphanumeric messages between mobile subscribers and external systems such as E-mail and voic mail systems. Messages can be no longer than 160 alpha-numeric characters and must contain no images or graphics. As wireless services evolved, Multimedia Messaging Service (MMS) was introduced and provided a way to send messages comprising a combination of text, sounds, images and video to MMS capable handsets.

Communication on wireless devices such as mobile phones and PDAs is frequently changing. Today you can use your wireless device to not only make phone calls, but to send and receive E-mail and IM. While you can use E-mail or IM for free if you have an Internet account, you will end up paying fees to you mobile carrier to use these services on a wireless device.

Technical Words and Phrases

account	[ə'kaunt]	n. 会计；账目；账户
adapter	[ə'dæptə(r)]	n. 改编者，适配器；多头电源插座
buddy	['bʌdɪ]	n. -ies 同伙，伙伴，朋友，搭档，好友
bundle	['bʌnd(ə)l]	n. 捆，束，包 v. 捆扎
chat	['tʃæt]	v. 聊天 n. 聊天
client	['klaɪənt]	n. 委托人，（律师的）当事人，接受社会照顾的人，顾客
employer	[ɪm'plɔɪə(r)]	n. 雇主，老板
instant	['ɪnst(ə)nt]	adj. 立即的，直接的，（食品）速溶的，方便的，即时的
spam	[spæm]	n. 垃圾邮件
transmission	[trænz'mɪʃ(ə)n]	n. 传动；传送，播送消息，动力传送器
videoconference	[vɪdɪəu'kɔnfərəns]	n. 视频会议
virus	['vaɪərəs]	n. 病毒

Multimedia Messaging Service (MMS)	多媒体信息服务
privacy issue	隐私泄漏
Short message service (SMS)	短消息服务
VoIP(Voice over Internet Protocol)	IP 电话

13.2 Reading Materials

13.2.1 Server

Server sometimes is defined as a computer or device on a network that manages network resources. For example, a file server is a computer and storage device dedicated to storing files. Any user on the network can store files on the server. A print server is a computer that manages one or more printers, and a network server is a computer that manages network traffic. A database server is a computer system that processes database queries.

Servers are often dedicated, meaning that they perform no other tasks besides their server tasks. On multiprocessing operating systems, however, a single computer can execute several programs at once. A server in this case could refer to the program that is managing resources rather than the entire computer.

From a hardware perspective, a server is simply a computer on your network that is configured to share its resources or run applications for the other computers on the network. You may have a server in place to handle file or database sharing between all users on your network, or have a server configured to allow all users to share a printer, rather than having a printer hooked up to each individual computer in your organization.

What makes the term server doubly confusing is that it can refer to both hardware and software. That is, it can be used to describe a specific software package running on a computer or the computer on which that software is running. The type of server and the software you would use depends on the type of network. LANs and WANs for example are going to use file and print servers while the Internet would use Web servers.

13.2.2 about Web Page

Have you ever wondered how a Web page (Fig 13.2) works? In order to talk about Web pages and how they work, you will want to understand four simple terms (and if some of these sounds like technical mumbo-jumbo the first time you read it, don't worry).

Web page - A Web page is a simple text file that contains not only text, but also a set of HTML tags that describe how the text should be formatted when a browser displays it on the screen. The tags are simple instructions that tell the browser to do things like change the font size or color, or arrange things in columns. The Web browser interprets these tags to decide how to format the text onto the screen.

Fig 13.2　web page

HTML - HTML stands for Hyper Text Markup Language. A "markup language" is a computer language that describes how a page should be formatted. If all you want to do is display a long string of black and white text with no formatting, then you don't need HTML. But if you want to change fonts, add colors, create headlines and embed graphics in your page, HTML is the language you use to do it.

Web browser - a Web browser, like Netscape Navigator or Microsoft Internet Explorer, is a computer program (also known as a software application, or simply an application) that does two things:

A Web browser knows how to go to a Web server on the Internet and request a page, so that the browser can pull the page through the network and into your machine (Fig 13.3).

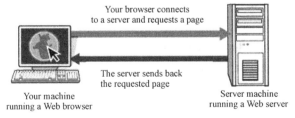

Fig 13.3　a browser connects to a server

A Web browser knows how to interpret the set of HTML tags within the page in order to display the page on your screen as the page's creator intended it to be viewed.

Web server - A Web server is combination of a computer and a piece of computer software, the software can respond to a browser's request for a page, and deliver the page to the Web browser through the Internet. You can think of a Web server as an apartment complex, with each apartment housing someone's Web page. In order to store your page in the complex, you need to pay rent on the space. Pages that live in this complex can be displayed to and viewed by anyone all over the world.

13.2.3　the Cell Approach

One of the most interesting things about a cell phone is that it is actually a radio.

The genius of the cellular system is the division of a city into small cells. (Cells are normally thought of as hexagons on a big hexagonal grid.) In a typical analog cell-phone system in the United States, the cell-phone carrier receives about 800 frequencies to use across the city. The carrier chops up the city into cells. Each cell is typically sized at about 10 square miles (26 square kilometers). This allows extensive frequency reuse across a city, so that millions of people can use cell phones simultaneously.

Each cell has a base station that consists of a tower and a small building containing the radio equipment (more on base stations later).

Cell phones have low-power transmitters in them. Many cell phones have two signal strengths: 0.6 watts and 3 watts. The base station is also transmitting at low power. Low-power transmitters have two advantages:

① The transmissions of a base station and the phones within its cell do not make it very far outside that cell. Therefore, the unconnected cells can reuse the same frequencies. The same frequencies can be reused extensively across the city.

② The power consumption of the cell phone, which is normally battery-operated, is relatively low. Low power means small batteries, and this is what has made handheld cellular phones possible.

The cellular approach requires a large number of base stations in a city of any size. A typical large city can have hundreds of towers. But because so many people are using cell phones, costs remain low per user.

13.2.4　Inside a Cell Phone

On a "complexity per cubic inch" scale, cell phones are some of the most intricate devices people play with on a daily basis. Modern digital cell phones can process millions of calculations per second in order to compress and decompress the voice stream.

If you take a cell phone apart, you find that it contains just a few individual parts:
- An amazing circuit board containing the brains of the phone
- An antenna
- A liquid crystal display (LCD)
- A keyboard (not unlike the one you find in a TV remote control)
- A microphone
- A speaker
- A battery

In the photos in Fig 13.4, you see several computer chips. Let's talk about what some of the individual chips do. The analog-to-digital and digital-to-analog conversion chips translate the outgoing audio signal from analog to digital and the incoming signal from digital back to analog. The digital signal processor (DSP) is a highly customized processor designed to perform signal-manipulation calculations at high speed.

Fig 13.4   the parts of a cell phone

The microprocessor handles all of the housekeeping chores for the keyboard and display, deals with command and control signaling with the base station and also coordinates the rest of the functions on the board.

The ROM and Flash memory chips provide storage for the phone's operating system and customizable features, such as the phone directory. The radio frequency (RF) and power section handles power management and recharging, and also deals with the hundreds of FM channels. Finally, the RF amplifiers handle signals traveling to and from the antenna.

Some phones store certain information in internal Flash memory, while others use external cards that are similar to SmartMedia cards.

Cell phones have such tiny speakers and microphones that it is incredible how well most of them reproduce sound. As you can see in the picture above, the speaker is about the size of a dime and the microphone is no larger than the watch battery beside it. Speaking of the watch battery, this is used by the cell phone's internal clock chip.

What is amazing is that all of that functionality - which only 30 years ago would have filled an entire floor of an office building - now fits into a package that sits comfortably in the palm of your hand!

13.2.5   Nokia N95

Combining a cell phone, GPS receiver, digital camera/camcorder, digital audio player, and PDA into an all-in-one multimedia computer/phone, the stylish Nokia N95 features a unique 2-way slide design (Fig 13.5) for easy switching between telephony, entertainment, and Web browsing. The quad-band Nokia N95 GSM/EDGE phone is also ready to run on 3G networks here in the US (850/1900 MHz UMTS/HSDPA), enabling fast downloads and streaming multimedia while on the go. It also includes integrated Wi-Fi connectivity (802.11b/g) for accessing open networks at work, at home, and on the road from a variety of wireless hotspots.

Fig 13.5   Nokia N95

The innovative 2-way slide concept makes it easy to switch between different modes, going from reading maps to watching a video with a simple slide. A numeric keypad slides out from one end of the device while dedicated media keys slide out from the opposite direction, converting the display into full screen landscape mode.

With the Carl Zeiss optics on the 5-megapixel camera, you can capture print quality photos and DVD-like quality video clips. The Nokia N95 actually has two cameras, a high resolution camera on the back of the device (the main camera in landscape mode) and a lower resolution camera on the front (CIF resolution, 352×288). The main camera on the back of the N95 supports an image capture resolution of 2592×1944 pixels. You can use both cameras to take still pictures and record videos. Images are saved as JPEG files while videos are recorded in the MPEG-4 file format with the .mp4 file extension, or in the file format with the .3gp file extension.

13.3 Knowledge about Translation
（翻译知识 13——否定形式）

英语与汉语在表达否定概念时所使用的语言手段有很大差别，因此在翻译时要特别加以注意。科技英语中常用否定语气的结构可分为全部否定、部分否定、双重否定和意义上的否定。

1. 全部否定

用 not 否定谓语动词是常见的一种全部否定形式。在翻译时一般都译成否定谓语，这与汉语的否定结构基本相同，除了 not 以外，其他表示全部否定意义的词有：no, nobody, none, nowhere, never, neither, nor, nothing，不管这些表示否定意义的词在句中作主语、宾语还是其他成分，这类句子通常译成否定句。

Nobody who has ever seen good quality color television can ever be completely happy with black and white again. 见过高质量彩色电视的人是不会再对黑白电视感到完全满意的。

2. 部分否定

英语中某些不定代词，如：all, every, both, 以及某些副词，如：always, often, quite, entirely, altogether 等与否定词连用时，表示的是部分否定。这种部分否定通常可译为"不全是"、"不都是"、"不常"、"未必都"、"并非完全"等。

Not all substances are conductors. 并非所有的物质都是导体。

The electrons within a conductor are **not** entirely free to move **but** are restrained by the attraction of the atoms among which they must move. 导体中的电子运动并不是完全自由的。电子必须在原子之间运动，从而要受到这些原子引力的束缚。

A programmer can avoid the use of assembly code in **all but** most demanding situations. 一个编程人员除了必须（用汇编语言）情况下可以避免使用汇编语言（编程）。

3. 双重否定

双重否定结构通常是由 no, not, never, nothing 等词与含有否定意义的词连用而构成的。这种结构形式上是否定，实质上是肯定，语气较强。翻译时可译为双重否定，有时也可译成肯定句。

You can do **nothing without** energy. 没有能量，你就什么也做不成。

In fact , there is **hardly** any sphere of life where electricity may **not** find useful application. 事实上，几乎任何一个生活领域都要用到电。

It is not uncommon for a programmable keyboard. 可编程的键盘现在也很通用。

后面这两句都是双重否定变成肯定语气。

4．意义上的否定

英语中有些词和词组在意义上表示否定。如：little（几乎没有），few（几乎没有），seldom（极少），scarcely（几乎不），hardly（很难，几乎不），too……to（太……以致不……），rather than（而不），fail to（不成功……，未能……）等。翻译时要译出否定的意义。

Metals, generally, offer **little** resistance and are good conductors. 通常金属几乎没有电阻，因而是良导体。

Glass conducts so little current that it is **hardly** measurable. 玻璃几乎不导电，因此很难测量其中的电流。

It may be easier to apply a force by pushing down **rather than** by pulling up. 向下推容易用力，向上举不容易用力。

5．否定转移

因英语与汉语在表达否定要领时所使用的词汇手段与语法手段都有很大的差别，所以在翻译时常用到两种转移，一是语法的转移。即否定主语或宾语转移成否定谓语，否定谓语转移成否定状语等。二是内容上的否定转移。即英语中的否定形式译成汉语时可用肯定形式，反之亦然。

No smaller quantity of electricity **than** the electron has ever been discovered. 从来没有发现过比电子电荷更小的电量。（由否定主语转移为否定谓语。）

Electric current **cannot** flow **easily** in some substances. 电流不能顺利地在某些物质中流动。（从逻辑上判断是否定状语 easily）。

Its importance **cannot** be stressed **too much**. 它的重要性怎么强调也不过分。（不要误译为：它的重要性不要强调的太过分。）

We have seen that the beta rays are **nothing but** a stream of electrons. 我们已经知道，β 射线只不过是一种电子流。

13.4 Exercises

1. Translate the Following Phrases into English

① 基于网络的通信　　　　　⑥ 传输媒体
②（本地）电话（服务）　　　⑦ 正在日益普及
③ 实时通信　　　　　　　　⑧ 被转换成数据包
④ 创建一种私人聊天室　　　⑨ 垃圾邮件
⑤ 与家人和朋友联系　　　　⑩ 免费软件和共享软件

2. Translate the Following Phrases into Chinese

① sounds a lot like regular telephone service
② online services
③ use an updated anti-virus program

④ scan your incoming and outgoing E-mail
⑤ participants at different sites
⑥ in front of the video camera
⑦ provide distance learning
⑧ alphanumeric messages
⑨ a global wireless service
⑩ LANs and WANs

3. Sentence Translation

① Of course, no technology is without a downside and Internet-based communications has plenty, such as viruses, privacy issues and spam.

② One of the fastest-growing forms of Internet communications is instant messaging, or IM.

③ For users who have free, or fixed-price Internet access, Internet telephony software essentially provides free telephone calls anywhere in the world.

④ Internet telephony consists of a combination of hardware and software that enables you to use the Internet as the transmission medium for telephone calls.

⑤ For this reason, it's important to use an updated anti-virus program, which will scan your incoming and outgoing E-mail for viruses.

⑥ In order for videoconference to work, the conference participants must use the same client or compatible software.

⑦ As wireless services evolved, Multimedia Messaging Service (MMS) was introduced and provided a way to send messages comprising a combination of text, sounds, images and video to MMS capable handsets.

⑧ Server sometimes is defined as a computer or device on a network that manages network resources.

⑨ What makes the term server doubly confusing is that it can refer to both hardware and software.

⑩ Each cell has a base station that consists of a tower and a small building containing the radio equipment (more on base stations later).

4. Translation

MP3, or MPEG Audio Layer III, is one method for compressing audio files. MPEG is the acronym for Moving Picture Experts Group, a group that has developed compression systems for video data, including that for DVD movies, HDTV broadcasts and digital satellite systems.

Using the MP3 compression system reduces the number of bytes in a song, while retaining sound that is near CD-quality. Anytime you compress a song, you will lose some of its quality, which is a trade off for being able to carry more music files in a smaller storage system. An MP3 player can be connected with a computer via the USB port and a smaller file size also allows the song to be downloaded from the Internet faster (see in Fig 13.6).

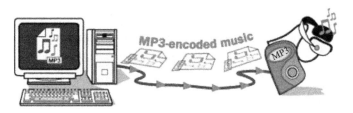

Fig 13.6　connected with a computer

13.5　课文参考译文

如果你上网，你可能是在用基于互联网的通信方式与家人、朋友或同事联系。互联网提供了很多联系方法：从给朋友发个即时消息、给同事寄个电子邮件，到打电话、视频会议。

互联网通信有很多优点，只要你已付了互联网的网络使用费（或你的老板付了），你就可以通过发送即时消息或用网络 IP 电话而省去打（普通）电话的费用。当然，没有一种技术是无缺点的，网络通信也有很多缺点，如病毒、泄密和垃圾邮件。

和所有技术（尤其是涉及互联网的技术）一样，在线通信的方法是在不断发展的。这里我们谈一下几种最主要的互联网通信方式。

13.5.1　即时消息（网上聊天）

互联网通信中发展最快的形式之一是即时消息（IM）（图 13.1），可把 IM 看成是在两个或更多人之间的计算机文字会议。IM 通信服务器可以创建个人聊天室，使你和另一人在网上实时聊天。一般 IM 系统还有提醒功能：当你的网上好友或联系人上网时提醒你，你就可以和某个人开始聊天。

当成百万个网络用户用 IM 与家人或朋友文字聊天时，IM 在工作中的应用也越来越普遍，当有事要谈时，公司的职员可以与不在一个办公室的经理或同事直接交谈，可以不再需要打电话。总之，IM 可以节省职员的时间并减少企业的通信开销。

13.5.2　网上通话和网络电话

网上通话由硬件和软件组成，使你可以用互联网作为电话的传输媒体，对已经免费或固定计费上网（如包月、包年）的用户，网络电话软件提供了可打给世界各地的免费电话。作为网络电话最简单的一种形式，计算机—计算机网络电话，只要在计算机上接一个话筒，把声音通过调制解调器送给与你有相匹配的网络电话软件的人。然而这种互联网网上通话也有它本身的问题，这种连接方式比用普通的电话慢，声音传输的质量也没普通电话好。

有许多种网上通话应用程序，有些如 CoolTalker（CoolTalker 是一个界面友好的网上即时联络工具，可以进行语音聊天和文字聊天）和 NetMeeting（网络会议）是与网络浏览器捆绑在一起的。也有一些是独立产品。网络电话产品有时称做 IP 电话、网上音频聊天或网上 IP 电话。

网上 IP 电话是另一种正日益普及的通信方式。网上 IP 电话是结合硬件和软件，用互联网传送打包的声音数据，而不是用传统的（称为公众电话网的）传输电路。待传输的声音先转换成打包的数据，然后就像一般的打包数据一样通过互联网或任何 IP 电话传输。当数据包到达目的地，再转换成声音数据供接收。你的电话是连接在一个网上 IP 电话适配器（硬件方面）。这个适配器与计算机主板相连接，电话通过互联网到一个标准的电话插口，与接收者的电话相连接。只有电话主叫方需要特殊的硬件（电话适配器）。

与上网要找一个网络服务提供商（ISP）一样，用网上 IP 电话也要找一个提供商，有些服务商会给出很多种付费方法，包括网内免费和网外按固定的次数或通话时间统一收费，可能打长途电话时要增加些费用等。尽管这些听起来与普通电话差不多，但要比普通电话便宜得多，因为你不必额外再付每个月的电话月租费。

13.5.3 电子邮件

简单说，电子邮件是在通信网上的传输信息。许多大型机、小型机及计算机网络都有电子邮件系统。有些电子邮件系统是局限于单个计算机系统或网络中的，但有些可通过网关送到其他计算机系统，把电子邮件送到世界的任何一个地方。

用一个电子邮箱（软件如微软的 Microsoft Outlook 或 Eudora），你可以写电子邮件，无论何地何人，只要你知道他的电子邮箱地址就可以给他发电子邮件。所有在线服务和网络服务提供商都提供电子邮件和网关服务。所以你可以和其他计算机系统的用户交换电子邮件。一封电子邮件一般只要几秒钟就可以送到目的地。电子邮件可以对一群人同时发信息或传送文件（群发），这是一种与一群人通信的特别有效的方式。

电子邮件最大的一个坏处是垃圾邮件。虽然按不同的定义，有时垃圾邮件是指送到千家万户的电子宣传品（通常是产品广告的电子邮件），但通常垃圾邮件是指散布特洛伊（一种病毒）和病毒的电子邮件。因此，用更新的杀毒软件扫描所收发电子邮件进行杀毒是很重要的。

13.5.4 视频会议

视频会议是指在两个或更多在不同地方的人通过用计算机网络传递声音和视频数据的方法在一起开会。每个参会者都有一个视频相机（摄像头）、话筒和喇叭与计算机相连接。当两个参会者相互讲话时，他们的声音通过网络送给对方，一方摄像头摄到的画面显示在另一方的计算机屏幕上。

开视频会议时，参会者必须用同样的客户软件或相匹配的软件。许多免费和共享的视频会议软件可从网上下载。许多网络相机（摄像头）（出售时）也捆绑视频软件。许多较新款的视频会议（软件）包还能与即时消息软件集成，用于多人和多形式的会议。

近年来，视频会议已成为教室中远距离通信的常用形式，用来提供远距离教学，远距离演讲和多校合作项目，是一种很有效的节约成本的方式。许多人认为视频会议提供了一种标准即时消息和电子邮件所不能做到的可见的联系和交流。

13.5.5 短消息服务和无线通信

短消息服务（SMS）是全球无线服务，使得文字消息可以在移动用户和其他外部系统如电子邮件和声音邮件系统之间传递。短消息内容不能超过 160 个文字或数字，不能包含图片。随着无线服务的发展，产生了多媒体信息服务（MMS），它可以传送包括文本、声音、图片和视频的信息给可接收 MMS 信息的手机。

基于无线设备如移动电话和 PDA（个人数据助理，类似具有手机功能的商务通）的通信方式在不断变化，今天你不但可以用无线设备打电话，还可以接收即时消息和电子邮件。如果你有一个网络账户，则电子邮件、即时消息都是免费的，在无线设备上用这些服务你不用给移动电话提供商付费。

13.6 阅读材料参考译文

13.6.1 服务器

服务器有时被定义成一台计算机或网络上的一个设备,用来管理网络资源,例如,文件服务器是一台用以存储文件的计算机及存储设备,网上的任何用户都可在服务器中存储文件。一台打印服务器是一台管理一个或多台打印机的计算机。一个网络服务器是管理网络事务的计算机。一个数据服务器是处理数据查询的计算机系统。

服务器经常是只处理服务方面的任务。但在多任务操作系统中,一台计算机同时可以运行多个程序,这种情况下服务器可能是指管理资源的程序,而不是整台计算机。

从硬件方面来看,一台服务器只是网络上的一台设置成资源可以共享或为网络上的其他计算机运行应用程序的计算机。在计算机网络中可以有一台服务器来处理所有用户共享的文件和数据,或有一个服务器设置(程序)使所有用户共享一台打印机,而不是网络中每台计算机分别接一台打印机。

服务器这个词有双重含义,既可指硬件,也可指软件,(这一点容易使人混淆)即它既可指在一台计算机上运行的软件包,也可指一台运行软件的计算机。网络的类型决定了服务器及其所用软件的类型。例如局域网和广域网用文件服务器和打印服务器,而互联网用网络服务器。

13.6.2 网页

你有没有好奇过,网页(如图 13.2 所示)是怎么回事?要了解网页和它们的运作过程,就要知道 4 个简单的词汇(如果你第一次看到这些词觉得比较难懂,别担心)。

网页——网页是有一个简单的文本文件,它不但含文本,而且还有一些 HTML 标志,表示当网页在浏览器屏幕上显示时文本的格式。这些标志是一些简单的指令,告诉浏览器去做些什么如改变字体或颜色、如何排列文本等。网页浏览器翻译这些标志并按要求在屏幕上放置这些文本。

HTML——HTML 表示超文本链接标志语言,标志语言是一种计算机语言,用来描述网页的格式。如果你只想显示一篇黑白文本可以不要格式,即你不用 HTML。但如果你想改变字体、颜色,创建网页标题和在页面中嵌入图片,HTML 就是你要用的语言。

网页浏览器——网页浏览器,如 Netscape Navigator(以开发 Internet 浏览器闻名的美国网景公司开发的浏览器)和微软的 Internet Explorer,是一个计算机程序(或称一个应用软件),它做这样两件事:

网页浏览器知道如何到达因特网上的网络服务器并申请一个网页,所以浏览器可以把网页信息通过网络取回来在你的电脑上显示出来(如图 13.3 所示)。

网页浏览器知道如何解读网页中的 HTML 标志,可以按网页制作者的要求把网页展示在你的屏幕上。

网络服务器——网络服务器是计算机和计算机软件的组合,这个计算机软件可以响应网页浏览器的请求,通过因特网发送网页给浏览器。你可以把网络服务器想象成一个公寓大楼,每个公寓房间中都装着一个人的网页。如果你也想把你的网页存储在这个大楼里,就需要租一个公寓。放在这个大楼里的网页可以送到世界各地显示,供人观看。

13.6.3 手机（蜂窝）技术

（关于手机最有意思的一件事是）手机实质上是一个无线接收机。

蜂窝系统是把一个城市划分成小的区域（蜂窝，蜂窝通常是正六边形构成的）。在美国一个典型的模拟手机蜂窝系统中，手机在穿过城市过程中可接收约 800 个频率，这些载波频率把城市分成很多区域，每个区域面积大约为 10 平方里（26 km^2）。这样很多频率在同一城市中（的不同区域）可以重复使用，所以数百万人可以同时打手机。

每个小区域中由一个发射塔和一幢含有无线电设备的建筑物（后来的基站有多幢建筑物），组成一个基站。

手机中有低功率的发射器，许多手机有两个信号（发射）强度：0.6 W 和 3 W，（手机）基站也是低功率发射。低功率发射（信号）有两个好处：

① 在区域中的基站和手机的（信号）发射使信号不会超出区域范围很远，所以不相邻的区域可以重复使用同样的频率。一个城市中同样的频率可以多次复用。

② 手机的功耗比较小，手机由电池供电，低功耗则电池也小一些，所以手机可做得比较小。

手机蜂窝技术需要在大小城市中建有大量基站，一个大城市有几百个发射塔，但因为有那么多的人用手机，所以对每个人来说用手机的费用不高。

13.6.4 手机的内部

按"每立方英寸复杂度"准则，手机是人们日常使用的最复杂的设备。为了压缩和解压声音流，现在数字手机每秒可以进行数百万次计算处理。

如果把手机拆开，可看到其中有这样一些部件：
- 含有手机的微处理器的电路主板
- 天线
- 液晶显示器（LCD）
- 键盘（与电视遥控器有点相似）
- 话筒
- 扬声器
- 电池

在图 13.4 的照片中，可看到一些（计算机）芯片，让我们看一下其中一些芯片。模-数和数-模转换芯片把外来的声音信号从模拟转换成数字并把接收到的信号从数字转换成模拟，数字信号处理器（DSP）是专门定制的，用来高速处理信号的计算。

微处理器处理键盘输入、显示器显示，处理来自基站的命令和控制信号以及与主板上其他功能相配合等所有管理任务。

ROM 和闪存芯片存储手机的操作系统和习惯设置，如电话目录。无线电频率（RF）和功率芯片处理功率管理和充电，并处理成百个调频频道，最后 RF 放大器芯片则处理天线接收和发送信号。

有些手机把一些信息存在内部的闪存芯片中，而另一些手机则用外部的类似 SM 卡的卡存储信息。

手机中的扬声器和话筒非常小，很难想象它们怎么能产生这么好的声音（效果）。扬声器只有（美元）一角硬币那么大，话筒则并不比它旁边的手表电池大。说到手表电池，它是用于手机内部的时钟电路的。

所有这些功能，30 年前（所有的设备）可能要占有一座办公楼的整整一层楼，现在都塞入一个手机，放在手掌中还足足有余，这是多么惊人啊。

13.6.5　诺基亚 N95

集手机、GPS 接收器、数码相机/摄像机、数字音乐播放器和个人掌上机于一身的诺基亚 N95 手机气度不凡，其带有双向滑盖（如图 13.5 所示）的设计可以很方便地在手机、娱乐和上网之间切换。方形的诺基亚 N95 GSM/EDGE 电话还可以接入 3G 网络，在美国是（850/1900 MHz UMTS/HSDPA），可以快速下载和浏览多媒体，它还集成了标准无线局域网（802.11b/g）模块，可以在工作场所、在家里或在路上（在有无线网络信号的地方）无线上网。

新颖的双向滑盖使它可以很方便地切换不同的工作模式，只要简单地一滑就可以从读地图切换到看视频，数字键盘从一端滑出，而专用媒体键从另一端滑出，并改变全屏显示的方式。

带有蔡司光学镜头和 500 万像素的相机，可以拍出高质量的照片和类似 DVD 的高质量视频。诺基亚 N95 实际上含有两个照相机，一个高分辨率相机（在后底板上，是全景模式的主相机）和一个较低分辨率相机（在前面板上，CIF 分辨率 352×288），N95 后底板上的主相机的分辨率达到 2592×1944 像素。这两个相机都可以拍静止照片和视频，照片以 JPEG 文件存储，视频以 MPEG-4 文件格式（扩展名为.mp4）或以 3GPP 文件格式（扩展名为.3gp）存储。

Unit 14　Electrical Appliances

> ***Pre-reading***
>
> *Read the following passage, paying attention to the question.*
> *1) What is the basic principle of a refrigerator?*
> *2) What is the function of refrigerant inside a refrigerator?*
> *3) What frequency radio wave the microwave oven usually used?*

14.1　Text

14.1.1　the Refrigerator

The basic idea behind a refrigerator is very simple: It uses the evaporation of a liquid to absorb heat. You probably know that when you put water on your skin it makes you feel cool. As the water evaporates, it absorbs heat, creating that cool feeling. Rubbing alcohol feels even cooler because it evaporates at a lower temperature. The liquid, or refrigerant, used in a refrigerator evaporates at an extremely low temperature, so it can create freezing temperatures inside the refrigerator. If you place your refrigerator's refrigerant on your skin (definitely NOT a good idea), it will freeze your skin as it evaporates.

There are five basic parts to any refrigerator (Fig 14.1):

- Compressor
- Heat-exchanging pipes - serpentine or coiled set of pipes outside the unit
- Expansion valve
- Heat-exchanging pipes - serpentine or coiled set of pipes inside the unit
- Refrigerant - liquid that evaporates inside the refrigerator to create the cold temperatures. Many industrial installations use pure ammonia as the refrigerant. Pure ammonia evaporates at -32 degrees Celsius.

The basic mechanism of a refrigerator works like this:

The compressor compresses the refrigerant gas. This raises the refrigerant's pressure and temperature, so the heat-exchanging coils outside the refrigerator allow the refrigerant to dissipate the heat of pressurization.

As it cools, the refrigerant condenses into liquid form and

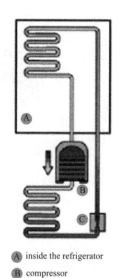

Ⓐ inside the refrigerator
Ⓑ compressor
Ⓒ expansion valve

Fig 14.1　parts of refrigerator

flows through the expansion valve.

When it flows through the expansion valve, the liquid refrigerant is allowed to move from a high-pressure zone to a low-pressure zone, so it expands and evaporates (light blue). In evaporating, it absorbs heat, making it cold.

The coils inside the refrigerator allow the refrigerant to absorb heat, making the inside of the refrigerator cold. The cycle then repeats.

Pure ammonia gas is highly toxic to people and would pose a threat if the refrigerator were to leak, so all home refrigerators don't use pure ammonia. You may have heard of refrigerants know as Freon a non-toxic replacement for ammonia. It has about the same boiling point as ammonia. However, Freon is not toxic to humans, so it is safe to use in your kitchen, but many large industrial refrigerators still use ammonia.

In the 1970s, it was discovered that the Freon then in use are harmful to the ozone layer, so as of the 1990s, all new refrigerators and air conditioners use refrigerants that are less harmful to the ozone layer.

14.1.2 the Air Conditioner

An air conditioner is basically a refrigerator without the insulated box. It uses the evaporation of a refrigerant, like Freon, to provide cooling. The mechanics of the Freon evaporation cycle are the same in a refrigerator as in an air conditioner.

① The compressor compresses cool Freon gas, causing it to become hot, high-pressure Freon gas.

② This hot gas runs through a set of coils so it can dissipate its heat, and it condenses into a liquid.

③ The Freon liquid runs through an expansion valve, and in the process it evaporates to become cold, low-pressure Freon gas.

④ This cold gas runs through a set of coils that allow the gas to absorb heat and cool down the air inside the building.

Mixed in with the Freon is a small amount of a lightweight oil. This oil lubricates the compressor.

Most air conditioners have their capacity rated in British thermal units (BTU). Generally speaking, a BTU is the amount of heat required to raise the temperature of one pound (0.45 kg) of water 1 degree Fahrenheit (0.56 degrees Celsius). Specifically, 1 BTU equals 1 055 joules.

The energy efficiency rating (EER) of an air conditioner is its BTU rating over its wattage. For example, if a 10 000-BTU air conditioner consumes 1 200 watts, its EER is 8.3 (10 000 BTU/1 200 watts). Obviously, you would like the EER to be as high as possible, but normally a higher EER is accompanied by a higher price.

14.1.3 the Microwave Oven

A microwave oven (Fig 14.2) consists of:
- a magnetron
- a magnetron control circuit (usually with a microcontroller)

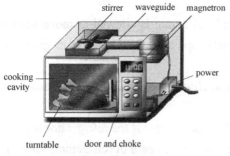

Fig 14.2 a microwave oven

- a waveguide
- a cooking chamber(or cooking cavity)

A microwave oven works by passing microwave radiation, usually at a frequency of 2 450 MHz (a wavelength of 12.24 cm), through the food. Water, fat, and sugar molecules in the food absorb energy from the microwave beam in a process called dielectric heating. Most molecules are electric dipoles, meaning that they have a positive charge at one end and a negative charge at the other, and is therefore twisted to and from as it tries to align itself with the alternating electric field induced by the microwave beam. This molecular movement creates heat. Microwave heating is most efficient on liquid water, and much less so on fats, sugars, and frozen water. Microwave heating is sometimes incorrectly explained as resonance of water molecules, but this occurs only at much higher frequencies, in the tens of gigahertz.

The cooking chamber itself is a Faraday cage enclosure to prevent the microwaves escaping into the surroundings. The oven door is usually a glass panel, but has a layer of conductive mesh to maintain the shielding. Since the mesh width is much less than the wavelength of 12 cm, the microwave radiation can not pass through the door, while visible light (with a much shorter wavelength) can.

Microwaves are radio waves. In the case of microwave ovens, the commonly used radio wave frequency is 2 450 MHz. Radio waves in this frequency range have an interesting property: they are absorbed by water, fats and sugars. When they are absorbed they are converted directly into atomic motion - heat. Microwaves in this frequency range have another interesting property: they are not absorbed by most plastics, glass or ceramics. Metal reflects microwaves, which is why metal pans do not work well in a microwave oven.

With wireless computer networks gaining in popularity, microwave interference has become a concern among those with wireless networks. Microwave ovens are capable of disrupting wireless network transmissions due to the fact that the microwave creates radio waves at about 2 450 MHz.

Technical Words and Phrases

absorb	[əb'sɔːb]	vt.	吸收，吸引
alcohol	['ælkəhɔl]	n.	酒精，乙醇，含酒精的饮料
ammonia	[ə'məunɪə]	n.	[化]氨，氨水
chamber	['tʃeɪmbə(r)]	n.	室，房间，议院，会所，（枪）膛
condense	[kən'dens]	v.	（使）浓缩，精简
dissipate	[dɪsɪpeɪt]	v.	驱散，（使）（云、雾、疑虑等）消散，浪费（金钱或时间）
evaporation	[ɪvæpə'reɪʃ(ə)n]	n.	蒸发，蒸发作用
Freon	['friːɔn]	n.	氟利昂
joule	[dʒuːl]	n.	[物]焦耳（功和能量的单位）

liquid	['lɪkwɪd]	n. 液体，流体，流音 adj. 液体的，清澈的，透明的
lubricate	['luːbrɪkeɪt]	vt. 润滑 v. 加润滑油
magnetron	['mæɡnɪtrɔn]	n. 磁电管，磁控管
molecule	['mɔlɪkjuːl]	n. 分子
refrigerant	[rɪ'frɪdʒərənt]	n. 致冷剂，冷却剂 adj. 致冷的，冷却的
refrigerator	[rɪ'frɪdʒəreɪtə(r)]	n. 冰箱
shield	['ʃiːld]	n. 防护物，护罩，盾，盾状物 vt. 保护，防护
toxic	['tɔksɪk]	adj. 毒的；中毒的，有毒的
twist	[twɪst]	n. 一扭，扭曲 vt. 拧，扭曲 vi. 扭弯，扭曲
valve	[vælv]	n. 阀；活门；气门
zone	[zəun]	n. 存储区；区段；区，带，层；区域，范围

14.2 Reading Materials

14.2.1 Gas and Propane Refrigerators

If you want to use a refrigerator where electricity is not available, you may use a gas or propane-powered refrigerator. These refrigerators are interesting because they have no moving parts and use gas or propane as their primary source of energy. Also, they use heat, in the form of burning propane, to produce the cold inside the refrigerator.

A gas refrigerator uses ammonia as the coolant, and it uses water, ammonia and hydrogen gas to create a continuous cycle for the ammonia. The refrigerator has five main parts:

● Generator - generates ammonia gas

● Separator - separates ammonia gas from water

● Condenser - where hot ammonia gas is cooled and condensed to create liquid ammonia

● Evaporator - where liquid ammonia evaporates to create cold temperatures inside the refrigerator

● Absorber - absorbs the ammonia gas in water

The cycle works like this:

Heat is applied to the generator. The heat comes from burning something like gas, propane or kerosene. In the generator is a solution of ammonia and water. The heat raises the temperature of the solution to the boiling point of the ammonia.

The boiling solution flows to the separator. In the separator, the water separates from the ammonia gas.

The ammonia gas flows upward to the condenser. The condenser is composed of metal coils and fins that allow the ammonia gas to dissipate its heat and condense into a liquid.

The liquid ammonia makes its way to the evaporator, where it mixes with hydrogen gas and evaporates, producing cold temperatures inside the refrigerator.

The ammonia and hydrogen gases flow to the absorber. Here, the water that has collected in the separator is mixed with the ammonia and hydrogen gases.

The ammonia forms a solution with the water and releases the hydrogen gas, which flows back to the evaporator. The ammonia-and-water solution flows toward the generator to repeat the cycle.

14.2.2 History about Microwave Oven

The idea for using microwaves to cook food was discovered by Percy Spencer who was working for Raytheon and was building magnetrons for radar sets. One day he was working on an active radar set when he had noticed a sudden and strange sensation, and saw that a chocolate bar he had in his pocket had melted. The holder of 120 patents, Spencer was no stranger to discovery and experiment, and realized what was happening.

The first food to be deliberately cooked with microwaves was popcorn, and the second was an egg (which exploded in the face of one of the experimenters).

In 1946 Raytheon patented the microwave cooking process and in 1947, they built the first commercial microwave oven, the Radarange. It was almost 6 feet (1.8 m) tall and weighed 750 pounds (340 kg). It was water-cooled and produced 3 000 watts, about three times the amount of radiation produced by microwave ovens today.

In the 1960s, Litton bought from Studebaker, Franklin Manufacturing assets, which had been manufacturing magnetrons, building, and selling microwave ovens similar to the Radarange.

Litton then developed a unique configuration of the microwave, the short, wide shape that is now common. The magnetron feed was also unique. The new oven was shown at a trade show in Chicago, and this was the beginning of the revolutionary growth of microwave ovens.

A number of other companies joined in the market, and for a time most systems were built by defense contractors, who were the most familiar with the magnetron. Litton was particularly well known in the restaurant business. By the late 1970s the technology had improved to the point where prices were falling rapidly. Formerly found only in large industrial applications, microwaves were increasingly becoming a standard fixture of most kitchens. The rapidly falling price of microprocessors also helped by adding electronic controls to make the ovens easier to use. By the late 1980s they were almost universal and currently it is estimated that nearly 95% of American households have a microwave.

14.2.3 Vacuum Cleaner

When you sip soda through a straw, you are utilizing the simplest of all suction mechanisms. Sucking the soda up causes a pressure drop between the bottom of the straw and the top of the straw. With greater fluid pressure at the bottom than the top, the soda is pushed up to your mouth.

This is the same basic mechanism at work in a vacuum cleaner, though the execution is a bit more complicated, it relies on a host of physical principles to clean effectively.

It may look like a complicated machine, but the conventional vacuum cleaner is actually made up of only six essential components (see in Fig 14.3):

- An intake port, which may include a variety of cleaning accessories
- An exhaust port
- An electric motor
- A fan

- A porous bag
- A housing that contains all the other components

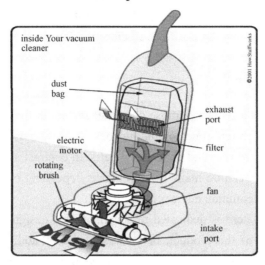

Fig 14.3   the vacuum cleaner

When you plug the vacuum cleaner in and turn it on, this is what happens:

The electric current operates the motor. The motor is attached to the fan, which has angled blades (like an airplane propeller).

As the fan blades turn, they force air forward, toward the exhaust port. When air particles are driven forward, the density of particles (and therefore the air pressure) increases in front of the fan and decreases behind the fan.

This pressure drop behind the fan is just like the pressure drop in the straw when you sip from your drink. The pressure level in the area behind the fan drops below the pressure level outside the vacuum cleaner (the ambient air pressure). This creates suction, a partial vacuum, inside the vacuum cleaner. The ambient air pushes itself into the vacuum cleaner through the intake port because the air pressure inside the vacuum cleaner is lower than the pressure outside.

As long as the fan is running and the passageway through the vacuum cleaner remains open, there is a constant stream of air moving through the intake port and out the exhaust port.

This stream of air acts just like a stream of water. The moving air particles rub against any loose dust or debris as they move, and if the debris is light enough and the suction is strong enough, the friction carries the material through the inside of the vacuum cleaner. This is the same principle that causes leaves and other debris to float down a stream. Some vacuum designs also have rotating brushes at the intake port, which kick dust and dirt loose from the carpet so it can be picked up by the air stream.

As the dirt-filled air makes its way to the exhaust port, it passes through the vacuum-cleaner bag. These bags are made of porous woven material (typically cloth or paper), which acts as an air filter. The tiny holes in the bag are large enough to let air particles pass by, but too small for most dirt particles to fit through. Thus, when the air current streams into the bag, all the air moves on through the material, but the dirt and debris collect in the bag.

You can put the vacuum-cleaner bag anywhere along the path between the intake tube and the exhaust port, as long as the air current flows through it. In upright vacuum cleaners, the bag is typically the last stop on the path. Immediately after it is filtered, the air flows back to the outside. In canister vacuums, the bag may be positioned before the fan, so the air is filtered as soon as it enters the vacuum.

14.2.4 Embedded System

An embedded system is a special-purpose computer system designed to perform one or a few dedicated functions, often with real-time computing constraints. It is usually embedded as part of a complete device including hardware and mechanical parts. In contrast, a general-purpose computer, such as a personal computer, can do many different tasks depending on programming. Embedded systems control many of the common devices in use today.

Since the embedded system is dedicated to specific tasks, design engineers can optimize it, reducing the size and cost of the product, or increasing the reliability and performance. Some embedded systems are mass-produced, benefiting from economies of scale.

Physically, embedded systems range from portable devices such as digital watches and MP3 players, to large stationary installations like traffic lights, factory controllers, or the systems controlling nuclear power plants. Complexity varies from low, with a single microcontroller chip, to very high with multiple units, peripherals and networks mounted inside a large enclosure.

In general, "embedded system" is not an exactly defined term, as many systems have some element of programmability. For example, Handheld computers share some elements with embedded systems - such as the operating systems and microprocessors which power them - but are not truly embedded systems, because they allow different applications to be loaded and peripherals to be connected.

14.3 Knowledge about Translation
（翻译知识 14——分离现象）

在一般情况下，句中的某些成分应当放在一起，如主语和谓语，动词和宾语等，但在科技英语中我们常可以看到这些成分离得较远，被其他成分隔开了，这种语言现象叫分离现象。起隔离作用的主要有：

① 各种短语：介词短语、分词短语、不定式短语等。
② 各种从句。
③ 句中的附加成分：插入语、同位语和独立成分。

分析这种隔离现象可以更好地帮助我们理解整个句子的原义。常见的分离现象有：

1. 主谓分离

The force that pushes you toward the front of the bus when it stops is the inertia of your body. 汽车停止时，推你往前倾的力就是你身体的惯性。

The keyboard, the most commonly used input device, is introduced in Unit 14. 在 14 单元中介绍了键盘，一种最常用的输入设备。

2. 动宾分离

有时作状语的介词短语等放在动词之后，用来修饰该动词，而把动词描述动作的对象——宾语隔开了。

But naysayers point out that tablet-style computers have been tested with flaws that vendors now claim to have overcome. 但是反对者指出，平板风格的计算机以前就试验过，但是有缺陷（虽然供应商声称已经克服了缺陷）。

overcome 的宾语是 that 或者说 flaws。

3. 复合谓语本身的分离

在复合谓语之间插入含有状语意义的介词短语或状语从句，使复合谓语本身产生分离现象。

Many forms of motion are highly complex, but they may **in all cases** be considered as being made up of translations and rotations. 很多运动的形式是非常复杂的，但在所有的情况下，都可以认为是由平移和转动合成的。

复合谓语 may be considered 之间插入 in all cases 造成分离现象。

4. 定语和被修饰名词的分离

An insulator is a material that offers a very high resistance to the passage **through it** of an electric current. 绝缘体是一种对通过它的电流产生很高阻力的材料。

of an electric current 是修饰 passage 的定语，被 through it 分隔开了。

5. 某些词与所要求介词的分离

The electric resistance of a wire is the ratio of the potential difference **between its two ends** to the current in the wire. 导线的电阻等于该导线两端之间的电位差与导线中电流的比值。

the ratio of A to B, 为"A"与"B"之比，这里被 between its two ends 隔开了。

14.4 Exercises

1. Put the Phrases into English

① 吸热
② 冰点温度
③ 热交换线圈
④ 扩张阀
⑤ 低压区
⑥ 空调
⑦ 压缩冷的氟里昂气体
⑧ 磁电管控制电路
⑨ 正电荷
⑩ 糖分子

2. Put the Phrases into Chinese

① the basic idea behind a refrigerator
② creat that cool feeling
③ evaporate at a lower temperature
④ liquid that evaporates inside the refrigerator
⑤ flow through the expansion valve
⑥ be highly toxic to people

⑦ run through a set of coils
⑧ absorb energy from the microwave beam
⑨ twist to and from
⑩ resonance of water molecules

3. Sentence Translation

① As the water evaporates, it absorbs heat, creating that cool feeling.

② If you place your refrigerator's refrigerant on your skin (definitely NOT a good idea), it will freeze your skin as it evaporates.

③ Pure ammonia gas is highly toxic to people and would pose a threat if the refrigerator were to leak.

④ The Freon liquid runs through an expansion valve, and in the process it evaporates to become cold, low-pressure Freon gas.

⑤ Microwave heating is most efficient on liquid water, and much less so on fats, sugars, and frozen water.

⑥ With wireless computer networks gaining in popularity, microwave interference has become a concern among those with wireless networks.

⑦ The first food to be deliberately cooked with microwaves was popcorn, and the second was an egg (which exploded in the face of one of the experimenters).

⑧ By the late 1970s the technology had improved to the point where prices were falling rapidly.

⑨ It may look like a complicated machine, but the conventional vacuum cleaner is actually made up of only six essential components.

⑩ You can put the vacuum - cleaner bag anywhere along the path between the intake tube and the exhaust port, as long as the air current flows through it.

4. Translation

A firewall is a protective system that lies, in essence, between your computer network and the Internet. When used correctly, a firewall prevents unauthorized use and access to your network. The job of a firewall is to carefully analyze data entering and exiting the network based on your configuration. It ignores information that comes from an unsecured, unknown or suspicious locations. A firewall plays an important role on any network as it provides a protective barrier against most forms of attack coming from the outside world.

14.5 课文参考译文

14.5.1 冰箱

冰箱的基本原理十分简单：用液体的蒸发来吸收热量。也许你知道当皮肤上有水时会感到凉快，这是因为水蒸发时，会吸收热量带来凉爽的感觉。如果擦酒精你会感到更凉爽，因为酒精的蒸发温度更低。冰箱中所用的液体或称制冷剂的蒸发温度特别低，所以可以在冰箱中产生冰冻的温度。如果你把冰箱中的制冷剂放在皮肤上（绝对不是好主意，别这样做），当它蒸发时会使皮肤结冰。

冰箱由5个基本部分组成（如图14.1所示）：
- 压缩机
- 热交换管——箱外的一组蜿蜒盘绕的管子
- 扩散阀
- 热交换管——箱内的一组蜿蜒盘绕的管子
- 制冷剂——在冰箱中蒸发以获得低温的液体，很多工业冰箱用纯氨作为制冷剂。纯氨的蒸发温度在 $-32\ ℃$

冰箱的基本工作原理如下：

压缩机压缩制冷剂气体，使制冷剂的压力和温度升高，冰箱外面的热交换管使制冷剂散热。当制冷剂冷却后，收缩成液态，流过扩散阀。

流过扩散阀时，液态制冷剂从高压管中流入低压管中，制冷剂膨胀蒸发成为（淡蓝色）气体，在蒸发过程中制冷剂要吸收热量，变成冷的气体。

冰箱内部的管子使制冷剂从冰箱中吸收热量，使冰箱内部变冷。如此反复循环。

纯氨气是有毒的，如果纯氨制冷剂泄漏对人是一种威胁，所以家用冰箱不用纯氨。你可能听说过Freon（氟里昂），一种氨的无毒替代品，与氨有基本相同的沸点，但氟里昂对人是没有毒的，所以可以安全地在厨房中使用，但许多大的工业冰箱仍用纯氨作制冷剂。

在（20世纪）70年代，发现当时所用的氟里昂破坏臭氧层，因此，到90年代，所有的冰箱和空调采用新制冷剂以减少对臭氧层的破坏作用。

14.5.2 空调

空调是没有隔热箱的冰箱，它利用制冷剂如氟里昂的蒸发进行制冷。空调和冰箱中的氟里昂的蒸发循环原理是相同的：

① 压缩机压缩冷的氟里昂气体，使它成为热的氟里昂高压气体。
② 热气体流过一组散热管，散热冷凝成液体。
③ 氟里昂液体流过扩散阀，蒸发成冷的低压氟里昂气体。
④ 冷的氟里昂气体流过一组管子，吸收热量，使室内温度降低。

有少量比较轻的油与氟里昂混合在一起，这种油对压缩机起润滑作用。

许多空调用英国热量单位（BTU）表示的其额定制冷（或制热）量。一般一个BTU是指一磅水（0.45 kg）升高一华氏度（0.56 ℃）所需吸收的热量。一个BTU等于1 055J。

空调的热效率（EER）是指空调的额定制冷（或制热）量除以它的（输入功率）瓦特，例如，一台额定制冷量为10 000 BTU的空调，额定消耗功率为1 200 W，则它的热效率为8.3。显然，你会要求热效率尽可能高，但一般热效率越高的空调价格越贵。

14.5.3 微波炉

微波炉由以下几部分组成（如图14.2所示）：
- 磁电管
- 磁电管的控制电路（一般带微处理器）
- 波导管
- 加热室

微波炉是通过微波（频率一般是2 450 MHz，波长为12.24 cm）穿透食品来加热的。食品中的水、脂肪和糖分子吸收微波中的能量，这个过程称为电介质加热。大部分分子是电双极性的，即分子的一端是正电荷，另一端是负电荷，在微波中，分子不断前后摆动，想与微波产生的交变的电场保持一致的方向。这种分子的运动就产生了热。微波加热对液态的水最有效，对脂肪、糖、结冰的水加热效果就差一些。有种不正确的解释说，微波加热是使水分子产生共振，实际上水分子共振所需的频率要高得多，要几十吉赫兹（GHz）（1 GHz=1 000 MHz）。

加热室本身是一个法拉第笼子（即利用法拉第电磁感应屏蔽原理制成），可以防止微波外泄，炉门通常是玻璃面板，但（玻璃中）有导电网格层起屏蔽作用。因为网格的宽度小于微波波长（12 cm），所以微波不能通过炉门（造成泄漏），但可见光（其波长比微波短得多）可以（通过炉门，所以你可以看到微波炉中的东西）。

微波是无线电波，微波炉常用的频率是 2 450 MHz。这个频率的无线电波有一个有趣的性质：这个波可以被水、脂肪和糖吸收，当它们吸收微波后，就把它直接转换成原子的运动——热量。在这个频率范围的微波还有另一个有趣的性质：大部分塑料、玻璃或陶器不能吸收微波，金属会反射微波，所以在微波炉中不能用金属盘。

随着无线计算机网络日益普及，微波干扰已成为无线网络中值得关注的问题。由于微波炉会发射出大约 2 450 MHz 的无线电波，可能会对无线网络的传输造成干扰。

14.6 阅读材料译文

14.6.1 汽油和丙烷冰箱

如果你想在没有电力供应的地方用冰箱，你可以用一台汽油或丙烷冰箱。有趣的是这些冰箱没有运动部件，用汽油或丙烷作为主要能源。它们也是以燃烧丙烷的形式产生热量，从而在冰箱内部制冷。

一个汽油冰箱用氨作冷却液，用水、氨和氢气产生氨的循环，这种冰箱有五个部分：
- 发生器——产生氨气体
- 分离器——从水中分离氨气体
- 冷凝器——热的氨气体在其中冷却凝结成液态
- 蒸发器——液态的氨气体在其中蒸发成气体，从而在冰箱内部产生低温
- 吸收器——吸收在水中的氨气

工作循环过程如下：

通过燃烧汽油、丙烷或煤油对发生器加热。在发生器中有氨水溶液，加热使溶液温度升高，达到氨水的沸点。

沸腾的溶液流过分离器，在分离器中氨气与水分开。

氨气往上进入到冷凝器中，冷凝器由细金属管组成，使氨气散热凝结成液体。

液态氨进入蒸发器，在蒸发器中与氢气混合并蒸发，使冰箱内部温度降低。

氨和氢气流入吸收器，在吸收器中，水收集氨和氢气并与它们混合。氨气与水形成氨溶液，并释放氢气，氢气流回蒸发器。氨水溶液流到发生器中再重复循环。

14.6.2 微波炉的由来

用微波炉来烹饪食品是 Percy Spencer 发现的。Percy Spencer 为 Raytheon 公司工作，制作雷达设备的磁电管。一天他正在一台开着的雷达设备前工作，突然他注意到一个奇怪的现象，他口袋里的巧克力熔化了。作为一个拥有 120 项专利的人，Spencer 对发明和实验并不陌

生，他马上意识到发生什么事了。

故意用微波去加热的第一种食品是玉米，第二种食品是一个鸡蛋（鸡蛋在一个实验者面前炸开了）。

1946 年 Raytheon 申请了微波烹饪方法的专利，1947 年，他们造出了第一台商业微波炉，称 Radarange，这台微波炉约有 6 英尺（1.8 m）高，重 750 磅（340 kg）。采用水冷方式，功率是 3 000 W，大概是现在微波炉功率的 3 倍。

1960 年，Litton 从 Studebaker 那儿购买了 Franklin 制造公司的资产，这个公司已经在制造磁电管，建造并销售类似于 Radarange 的微波炉。

然后 Litton 就开发了一种独特的微波炉，就是现在通用的短而宽的微波炉。所用的磁电管也是独特的。新的微波炉在芝加哥的一个贸易展销会上展出，这是微波炉迅速发展的开始。

许多其他公司也加入到这个市场中来了，一时间，国防产品的承包商制造了很多（微波炉）系统，因为他们对磁电管最熟悉。Litton 反而是在旅馆业最出名（的人）。到 1970 年后期，微波炉制造技术已得到很大的改进，价格迅速下降。以前的微波炉只在工业中得到应用，后来微波炉很快成为许多厨房的必备电器。迅速下降的价格以及附加的电子控制使微波炉使用起来很方便，所以到 20 世纪 80 年代后期，微波炉变得十分普及，现在据估计约有 95% 的美国家庭拥有微波炉。

14.6.3 真空吸尘器

当你用一根吸管吸汽水时，你就是在用最简单的抽水泵原理，吸汽水引起吸管中的压力下降，瓶子底部的液体压力比顶部大，汽水就被压入你的嘴中。

真空吸尘器最基本的工作原理与此相同，虽然工作过程稍复杂些，它利用一些物理原理使吸尘器的效率更高。

也许看起来像一个复杂的机器,但传统的真空吸尘器实际上仅由 6 个基本部件组成的（如图 14.3 所示）：

- 一个进口，也许还包括各种（形状的）清洁附件
- 一个排气孔
- 一个电动机
- 一个风扇
- 一个多孔的袋子（垃圾袋）
- 一个（盛放所有其他部件的）外壳

当插上真空吸尘器的电源插头并打开开关，就开始吸尘了：

电流使电动机旋转，电动机与按一定角度放置叶片（像飞机上的螺旋桨）的风扇连在一起。

当风扇叶片转动时，它们推动空气向前经过排气口，当空气微粒被推着向前时，风扇前面的空气密度增加（导致空气压力增加），风扇后面的空气密度减少（导致空气压力减少）。

风扇后面的空气压力减少就像是你吸汽水时在吸管中的压力减少。风扇后面的压力小于真空吸尘器外面的压力（周围的空气压力），这就在真空吸尘器中产生（局部真空）吸力，因为真空吸尘器内部的空气压力低于外部的压力，周围的空气压入真空吸尘器的进口。

只要风扇在旋转，并保持真空吸尘器的出入口畅通，就有一个不断的空气流从进口流进，从排气口流出。

这个气流就像一个水流，当空气运动时与松动的灰尘及碎屑互相摩擦，如果碎屑足够轻，

吸力足够强，摩擦力带着它们进入吸尘器，就和水流流下来时带着树叶和其他碎片一样。有的真空吸尘器在进口处还设计了一个旋转的刷子，它把灰尘和垃圾从地毯上扫起来（使它们松动可以让空气流吸起）。

当充满灰尘的空气通过真空吸尘器到达排风口时，它经过真空吸尘器的垃圾袋，垃圾袋是用有微孔的织物（如布或纸）做的，它的作用像一个空气过滤器，袋子的微孔可让空气流过，但垃圾不能穿过。因此当空气流进入袋子，所有的空气流出去了，但垃圾和碎屑留在袋子里。

可以把真空吸尘器的垃圾袋放在进风口和排气口之间任何空气流可以经过的地方。在立式真空吸尘器中，袋子一般放在排风口前，空气流经它过滤后就马上排到外面去了。在罐状（卧式）真空吸尘器中，袋子可放在风扇前，空气流一进入吸尘器就先过滤。

14.6.4 嵌入式系统

嵌入式系统是一种为实现一种或几种功能而设计的专用计算机系统，经常用做计算机实时控制，嵌入式系统通常作为一个含有硬件和机械部件的完整设备的一个（嵌入）部分。作为对照，通用型计算机如个人计算机则根据所编程序可以做很多种不同的工作。目前很多常用设备中都有嵌入式系统。

因为嵌入式系统只要求完成指定的任务，所以设计工程师可以优化这个系统，减小其尺寸，降低成本，或增加其可靠性和性能。有些嵌入式系统是批量生产的，有一定的规模经济效益。

外观上，嵌入式系统的种类很多，从便携式设备如数字钟、MP3播放器到很大的装置中，如交通灯、工厂的控制器或核电站的控制系统。系统的复杂度变化范围也很大，从较简单的单一微处理器芯片到非常复杂的带有多个单元、外设和网络的系统，被安装在一个大的设备中。

一般来说，"嵌入式系统"这个词定义并不准确，许多系统还有一些可编程的元件，例如手提电脑与嵌入式系统有些元素是共享的——如操作系统和控制它们的微处理器，但手提电脑并不是真正的嵌入式系统，因为手提电脑可以外接不同的负载和外设，使之有不同的应用。

Unit 15　I/O Devices

> **Pre-reading**
>
> *Read the following passage, paying attention to the question.*
> *1) What is a keyboard?*
> *2) What is the function of a mouse?*
> *3) What help Inkjet Printer retain their advantage in the realm of colour once a time?*
> *4) What is the primary principle at work in a laser printer?*

15.1　Text

15.1.1　Keyboards

At its essence, a keyboard is a series of switches connected to a microprocessor that monitors the state of each switch and initiates a specific response to a change in that state.

Two types of switch are commonly used: mechanical and rubber membrane. Mechanical switches are simply spring-loaded, so when pressed down they complete the circuit and then break it again when released.

Membranes are composed of three sheets: the first has conductive tracks printed on it, the second is a separator with holes in it and the third is a conductive layer with bumps on it. A rubber mat over this gives the springy feel. When a key is pressed, it pushes the two conductive layers together to complete the circuit. On top is a plastic housing, which includes sliders to keep the keys aligned.

An important factor for keys is their force displacement curve, which shows how much force is needed to depress a key, and how this force varies during the key's downward travel. Research shows most people prefer 80 g to 100 g, but games consoles may go to 120 g or higher.

The keys are connected up as a matrix, and their row and column signals feed into the keyboard's own microcontroller chip. This is mounted on a circuit board inside the keyboard, and interprets the signals with its built-in firmware program. A particular key press might signal as row 3, column B, so the controller might decode this as an A and send the appropriate code for A back to the PC. These "scan codes" are defined as standard in the PC's BIOS, though the row and column definitions are specific only to that particular keyboard.

Increasingly, keyboard firmware is becoming more complex as manufacturers make their keyboards more sophisticated. It is not uncommon for a programmable keyboard, in which some keys have switchable multiple functions, to need 8 KB of ROM to store its firmware. Most programmable functions are executed through a driver running on the PC.

Keyboards have changed very little in layout since their introduction. In fact, the most common change has simply been the natural evolution of adding more keys that provide additional functionality. A typical keyboard has four basic types of keys (Fig 15.1):

- Typing keys
- Numeric keypad
- Function keys
- Control keys

The typing keys are the section of the keyboard that contains the letter keys, generally laid out in the same style that was common for typewriters. The numeric keypad is a part of the natural evolution. As the use of computers in business environments increased, so did the need for speedy data entry. Since a large part of the data was numbers, a set of 17 keys was added to the keyboard. These keys are laid out in the same configuration used by most adding machines and calculators, to facilitate the transition to computer for clerks accustomed to these other machines.

In 1986, IBM extended the basic keyboard with the addition of function and control keys. The function keys, arranged in a line across the top of the keyboard, could be assigned specific commands by the current application or the operating system. Control keys provided cursor and screen control. Four keys arranged in an inverted T formation between the typing keys and numeric keypad allows the user to move the cursor on the display in small increments. Common control keys include: Home, End, Insert, Delete etc.

15.1.2 Mice (or Mouse)

In the early 1980s the first PCs were equipped with the traditional user input device - a keyboard. By the end of the decade however, a mouse device had become an essential for PCs running the GUI-based Windows operating system.

The commonest mouse used first is (opto-electronic) mechanical mouse. Its ball is steel for weight and rubber-coated for grip, and as it rotates it drives two rollers, one each for x and y displacement.

These rollers then turn two disks with radial slots cut in them. Each disk rotates between a photo-detector cell, and each cell contains two offset light emitting diodes (LEDs) and light sensors. As the disk turns, the sensors see the light appear to flash, showing movement, while the offset between the two light sensors shows the direction of movement (see in Fig 15.2).

Fig 15.1 the keyboard

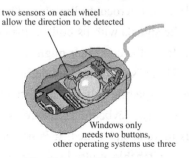

Fig 15.2 inside the mouse

Also inside the mouse are a switch for each button, and a microcontroller, which interpret the signals from the sensors and the switches, using its firmware program to translate them into packets of data that are sent to the PC. Serial mice use voltages of 12 V and an asynchronous protocol from Microsoft comprised of three bytes per packet to report x and y movement plus button presses. PS/2 mice and USB mice use 5 V and an IBM-developed communications protocol and interface.

1999 saw the introduction of the most radical mouse design advancement. Gone are the mouse ball and other moving parts inside the mouse used to track the mouse's mechanical movement, replaced by a tiny complementary metal oxide semiconductor (CMOS) optical sensor - the same chip used in digital cameras - and an on-board digital signal processor (DSP).

Here's how the sensor and other parts of an optical mouse work together:

The CMOS sensor sends each image to a digital signal processor (DSP) for analysis. The DSP detects patterns in the images and examines how the patterns have moved since the previous image. Based on the change in patterns over a sequence of images, the DSP determines how far the mouse has moved and sends the corresponding coordinates to the computer. The computer moves the cursor on the screen based on the coordinates received from the mouse. This happens hundreds of times each second, making the cursor appear to move very smoothly.

Optical mice have several benefits over track-ball mice:

- No moving parts means less wear and a lower chance of failure.
- There's no way for dirt to get inside the mouse and interfere with the tracking sensors.
- Increased tracking resolution means a smoother response.
- They don't require a special surface, such as a mouse pad.

15.1.3 Inkjet Printers

Traditionally, inkjets have had one massive attraction over laser printers; their ability to produce colour, and that is what makes them so popular with home users. Since the late 1990s, when the price of colour laser printers began to reach levels, which made them viable for home users, this advantage has been less definitive. However, in that time the development of inkjets capable of photographic-quality output has done much to help them retain their advantage in the realm of colour.

An inkjet printer is any printer that places extremely small droplets of ink onto paper to create an image. If you ever look at a piece of paper that has come out of an inkjet printer, you know that:

- The dots are extremely small (usually between 50 and 60 microns in diameter), so small that they are tinier than the diameter of a human hair (70 microns)!
- The dots are positioned very precisely, with resolutions of up to 1440×720 dots per inch (dpi).
- The dots can have different colors combined together to create photo-quality images.

Main parts of a typical inkjet printer include:

- Print head - the core of an inkjet printer, the print head contains a series of nozzles that are used to spray drops of ink.
- Ink cartridges - depending on the manufacturer and model of the printer, ink cartridges come in various combinations.

- Print head stepper motor - a stepper motor moves the print head assembly (print head and ink cartridges) back and forth across the paper.
- Paper tray/feeder - most inkjet printers have a tray that you load the paper into. Some printers dispense with the standard tray for a feeder instead.
- Paper feed stepper motor - this stepper motor powers a roller to move the paper in the exact increment needed to ensure a continuous image is printed.
- Control circuitry - a small but sophisticated amount of circuitry is built into the printer to control all the mechanical aspects of operation, as well as decode the information sent to the printer from the computer.

15.1.4 the Laser Printer

The laser printer was introduced by Hewlett-packard in 1984, based on technology developed by Canon. It worked in a similar way to a photocopier, the difference being the light source. With a photocopier a page is scanned with a bright light, while with a laser printer the light source is, not surprisingly, a laser.

The primary principle at work in a laser printer is static electricity, the same energy that makes clothes in the dryer stick together or a lightning bolt travel from a thundercloud to the ground. Static electricity is simply an electrical charge built up on an insulated object, such as a balloon or your body. Since oppositely charged atoms are attracted to each other, objects with opposite static electricity fields cling together.

A laser printer uses this phenomenon as a sort of "temporary glue." The core component of this system is the photoreceptor, typically a revolving drum or cylinder. This drum assembly is made out of highly photoconductive material that is discharged by light photons.

Initially, the drum is given a total positive charge by the charge corona wire, a wire with an electrical current running through it. (Some printers use a charged roller instead of a corona wire, but the principle is the same.) As the drum revolves, the printer shines a tiny laser beam across the surface to discharge certain points. In this way, the laser "draws" the letters and images to be printed as a pattern of electrical charges - an electrostatic image. The system can also work with the charges reversed - that is, a positive electrostatic image on a negative background.

After the pattern is set, the printer coats the drum with positively charged toner - a fine, black powder. Since it has a positive charge, the toner clings to the negative discharged areas of the drum, but not to the positively charged "background." This is something like writing on a soda can with glue and then rolling it over some flour: The flour only sticks to the glue-coated part of the can, so you end up with a message written in powder.

With the powder pattern affixed, the drum rolls over a sheet of paper (Fig 15.3), which is moving along a belt below. Before the paper rolls under the drum, it is given a negative charge by the transfer corona wire (charged roller). This charge is stronger than the negative charge of the electrostatic image, so the paper can pull the toner powder away. Since it is moving at the same speed as the drum, the paper picks up the image pattern exactly.

Finally, the printer passes the paper through the fuser (Fig 15.4), a pair of heated rollers. As the paper passes through these rollers, the loose toner powder melts, fusing with the fibers in the paper. The fuser rolls the paper to the output tray, and you have your finished page. The fuser also heats up the paper itself, of course, which is why pages are always hot when they come out of a laser printer or photocopier.

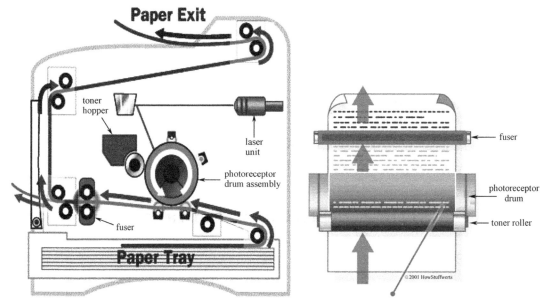

Fig 15.3   the path of a piece of paper through a laser printer   Fig 15.4   printer passes the paper

After depositing toner on the paper, the drum surface passes the discharge lamp. This bright light exposes the entire photoreceptor surface, erasing the electrical image. The drum surface then passes the charge corona wire, which reapplies the positive charge.

Technical Words and Phrases

Affix	[əˈfɪks]	vt. 使固定；贴上；粘上
array	[əˈreɪ]	n. 排列，编队，军队，衣服，大批 vt. 部署，穿着，排列
cartridge	[ˈkɑːtrɪdʒ]	n. 墨粉盒，盒式磁盘[带](机)；夹头
clerk	[ˈklɑːk]	n. 书记；秘书；文书，记录员；书记员，售货员；店员
column	[ˈkɔləm]	n. 列，圆柱，柱状物，专栏，纵队
corona	[kəˈrəunə]	n. 冠状物，王冠，光环
displacement	[dɪsˈpleɪsmənt]	n. 移置，转移，取代，置换，位移，排水量
drum	[drʌm]	n. 鼓，鼓声，鼓状物
facilitate	[fəˈsɪlɪteɪt]	vt.（不以人作主语的）使容易，使便利，推动
firmware	[ˈfɜːmweə(r)]	n. [计] 固件，固化（在硬件中的）程序
fuser	[ˈfjuːz]	n. 熔辊
keyboard	[ˈkiːbɔːd]	n. 键盘
matrix	[ˈmeɪtrɪks]	n. 矩阵，矩[方，点]阵，真值表
membrane	[ˈmembreɪn]	n. 薄膜，隔膜，细胞膜

mouse	[maʊs]	n. 鼠标（原义为老鼠，复数为 mice）
opto-electronic	[ɔptə-ɪlek'trɒnɪk]	adj. 光电的
phenomenon	[fɪ'nɒmɪnən]	n. 现象
powder	['paʊdə(r)]	n. 粉，粉末，火药，尘土 vt. 搽粉于，撒粉，使成粉末
protocol	['prəʊtəkɒl]	n. 草案，协议
roller	['rəʊlə(r)]	n. 滚筒，辊，卷
row	[rəʊ]	n. 排，行 v. 划（船）
shine	[ʃaɪn]	v. 照耀，发光 vt. 擦亮 n. 光泽，光亮
viable	[vaɪəb(ə)l]	adj. 可行的；可实施的

inkjet printer	喷墨打印机
print head stepper motor	喷头步进电机
Laser Printer	激光打印机

15.2 Reading Materials

15.2.1 Touchscreen

A touchscreen is an intuitive computer input device that works by simply touching the display screen, either by a finger, or with a stylus, rather than typing on a keyboard or pointing with a mouse.

The touchscreen interface - whereby users navigate a computer system by touching icons or links on the screen itself - is the most simple, intuitive, and easiest to learn of all PC input devices and is fast becoming the interface of choice for a wide variety of applications, such as:

Public Information Systems: Information kiosks, tourism displays, and other electronic displays are used by many people that have little or no computing experience. The user-friendly touchscreen interface can be less intimidating and easier to use than other input devices, especially for novice users, making information accessible to the widest possible audience.

Restaurant/POS Systems: Time is money, especially in a fast paced restaurant or retail environment. Because touchscreen systems are easy to use, overall training time for new employees can be reduced. And work can get done faster, because employees can simply touch the screen to perform tasks, rather than entering complex keystrokes or commands.

Customer Self-Service: In today's fast pace world, waiting in line is one of the things that have yet to speed up. Self-service touchscreen terminals can be used to improve customer service at busy stores, fast service restaurants, transportation hubs, and more.

Control/Automation Systems: The touchscreen interface is useful in systems ranging from industrial process control to home automation. By integrating the input device with the display, valuable workspace can be saved. And with a graphical interface, operators can monitor and control complex operations in real-time by simply touching the screen.

Any touchscreen system comprises the following three basic components.

A touchscreen sensor panel, that sits above the display and which generates appropriate voltages according to where, precisely, it is touched.

A touchscreen controller, that processes the signals received from the sensor and translates these into touch event data which is passed to the PC's processor, usually via a serial or USB interface.

A software driver provides an interface to the PC's operating system and which translates the touch event data into mouse events, essentially enabling the sensor panel to "emulate" a mouse.

There are several types of touchscreens; Here we discuss the infrared touchscreens (Fig 15.5). Infrared touchscreens are based on light-beam interruption technology. Instead of placing a layer on the display surface, a frame surrounds it. The frame assembly is comprised of printed wiring boards on which the opto-electronics are mounted and is concealed behind an IR-transparent bezel.

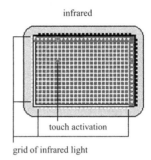

Fig 15.5   infrared touchscreen

The frame contains light sources - or light-emitting diodes - on one side, and light detectors - or photosensors - on the opposite side. The effect of this is to create an optical grid across the screen. When any object touches the screen, the invisible light beam is interrupted, causing a drop in the signal received by the photosensors. Based on which photosensors stop receiving the light signals, it is easy to isolate a screen coordinate.

Infrared touch systems are solid state technology and have no moving mechanical parts. As such, they have no physical sensor that can be abraded or worn out with heavy use over time. Furthermore, since they do not require an overlay - which can be broken - they are less vulnerable to vandalism and also extremely tolerant of shock and vibration.

15.2.2   Colour Lasers

Initially, most commercial laser printers were limited to monochrome printing (black writing on white paper). But now, there are lots of color laser printers on the market.

Essentially, color printers work the same way as monochrome printers, except they go through the entire printing process four times - one pass each for cyan (blue), magenta (red), yellow and black. By combining these four colors of toner in varying proportions, you can generate the full spectrum of color.

There are several different ways of doing this. Some models have four toner (Fig 15.6) and developer units on a rotating wheel. The printer lays down the electrostatic image for one color and puts that toner unit into position. It then applies this color to the paper and goes through the process again for the next color. Some printers add all four colors to a plate before placing the image on paper. Some more expensive printers actually have a complete printer unit - a laser assembly, a drum and a toner system - for each color. The paper simply moves past the different drumheads, collecting all the colors in a sort of assembly line.

Most modern laser printers have a native resolution of 600 or 1 200 dpi. Rated print speeds vary between 3 and 5 ppm in colour and 12 to 14 ppm in monochrome. A key area of development, pioneered by Lexmark's 12 ppm LED printer launched in the autumn of 1998, is to boost colour print speed up to the same level as mono with simultaneous processing of the four toners and one-pass printing.

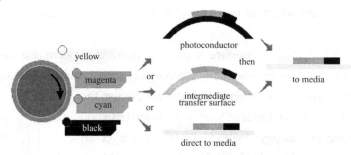

Fig 15.6   four colors of toner in a color laser printer

Apart from their speed, one of the main advantages of colour lasers is the durability of their output - a function of the chemically inert toners that are fused onto the paper's surface rather than absorbed into it, as with most inkjets. This allows colour lasers to print well on a variety of media, without the problems of smudging and fading that beset many inkjets. Furthermore, by controlling the amount of heat and pressure in the fusing process, output can be given a user-controllable "finish", from matte through to gloss.

15.2.3   Electronic Circuit Simulation

Electronic circuit simulation utilizes mathematical models to replicate the behavior of an actual electronic device or circuit. Simulating a circuit's behavior before actually building it greatly improves efficiency and provides insights into the behavior of electronics circuit designs. In particular, for integrated circuits, the tooling (photomasks) is expensive, breadboards are impractical, and probing the behavior of internal signals is extremely difficult. Therefore almost all IC design relies heavily on simulation. The most well known analog simulator is SPICE. Probably the best known digital simulators are those based on Verilog and VHDL.

Some electronics simulators integrate a schematic editor, a simulation engine, and on-screen waveforms and make "what-if" scenarios easy and instant. They also typically contain extensive model and device libraries. These models typically include IC specific transistor models, generic components such as resistors, capacitors, inductors and transformers, user defined models. Printed circuit board (PCB) design requires specific models as well, such as transmission lines for the traces and IBIS models for driving and receiving electronics.

While there are strictly analog electronics circuit simulators, popular simulators often include both analog and digital simulation capabilities, and are known as mixed-mode simulators. This means that any simulation may contain components that are analog, event driven (digital or sampled-data), or a combination of both. An entire mixed signal analysis can be driven from one integrated schematic. All the digital models in mixed-mode simulators provide accurate specification of propagation time and rise/fall time delays.

Here is an example of mixed-mode simulator_ Electronics Workbench. Electronics Workbench is a design tool that provides you with all the components and instruments necessary to create board-level designs. Fig 15.7 is the interface of EWB. It has complete mixed analog and digital simulation and graphical waveform analysis, allowing you to design your circuit and then analyze it using different simulated instruments and analysis options. It is fully integrated and interactive, thus you can change your circuits quickly, allowing fast and repeated what-if analysis.

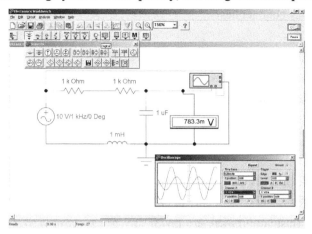

Fig 15.7   the interface of EWB

The simulator in Electronics Workbench, like other general-purpose simulators, has four main stages: input, setup, analysis and output.

● At the input stage, after you have built schematic, assigned values and chosen an analysis, the simulator reads information about your circuit.

● At the setup stage, the simulator constructs and checks a set of data structures that contain a complete description of your circuit.

● At the analysis stage, the circuit analysis specified in the input stage is performed. This stage occupies most of CPU execution time and actually is the core of circuit simulation. The analysis stage formulates and solves circuit equations for the specified analyses and provides all the data for direct output or post-processing.

● At the output stage, you view the simulation results. You can view results on instruments such as the oscilloscope, or on graphs that appear when you run an analysis from the Analysis menu or when you choose Analysis/Display Graphs.

15.3 Knowledge about Translation
（翻译知识15——省略和插入语）

英语的句子中除主、动、宾及定语、状语及补足语外，常有同位语、插入语及独立定语，这些句子的附加成分在语法中虽然不处于重要的地位，但对阅读科技专业书刊往往会起一些干扰作用，妨碍我们对句子中主要成分的理解和分析。

1. 省略

英语句子中常省略一些成分使语言精练，但有时这种省略给我们的阅读和翻译带来一些困难，现在我们来看一下在科技英语中常见的省略情况。

① 并列结构中某些相同成分的省略。

在用 and ,but ,or ,while, so 等词汇连起来的并列句中，常省去与前面相同的成分。

Stone, brick and glass conduct heat poorly compared to metals, **but** well compared to wood, paper, cloth and air. 石头、砖和玻璃与金属相比导热性很差，但与木头、纸、布和空气相比，**则导热性**较好。

but 后面省去了 they conduct heat。

② 状语从句中省略。

在一些状语从句中，如果其主语与主句中的主语相同，且谓语中含有动词 be，则从句中的主语和 be 常常可以省略。

Potential energy, though not so obvious as kinetic energy, exists in many things. 位能（势能）虽不如动能那样明显，但很多物体都具有势能。

though 后面省略了 it is。

③ 常用省略。

有些省略已成为习惯用法，如：if possible (可能的话)，if necessary（必要的话），if any(如果主句含有否定的意思，译成：即使有，也很少，如果主句是肯定的，译成：若有……。

In non-conductors (insulators) there are few **if any** free electrons. 在绝缘体中几乎没有自由电子，**即使有，也极少**。

这里 if any 也可以不译。

There are very few mistakes in his work, if any. 他工作极少出错。

④ 比较句中的省略。

The hotter the gas, the faster do the molecules move. 气体越热，分子就运动得越快。

gas 后面省略了 is。

In a series circuit the total resistance is the sum of the individual resistances. The current in each unit is the same as in all other units. 在串联电路中，总电阻等于各个电阻的总和，通过每一个元件的电流与通过其他元件的电流相同。

as 后面省略了 the current is。

2．插入语

插入语也是英语中常有的现象，它也能造成句子的分离现象，在阅读和翻译时把它先放在一边，免得对句子的主要意思产生干扰。一般插入语会用逗号断开，所以还是比较好判断的。

① 插入语中见到最多的是同位语，如果名词后面有另一名词（或代词）、词组或句子对该名词作进一步的说明，而这两者在语法上又处于同等的位置，则称为同位语。同位语和前面的名词之间可用标点隔开也可以不隔开。有时用 such as, as, or, that is, for example, particularly（特别是），including（包括）等引出同位语。

Color television breaks down the pictures into three colors - **red, green, and blue**. 彩色电视把图像分解为三种颜色——红、绿、蓝。

On October 17, 1831, Faraday succeeded in making one of the greatest discoveries in the field of electricity—**the electromagnetic induction**. 法拉第在 1831 年 10 月 17 日成功地做出了电学领域中的一项最伟大的发现——电磁感应。

Their (electrons) orbits, or paths, form concentric shells, or layers, some what like the layers of an onion. 电子的轨道（即轨迹）形成一些同心的外壳（即同心层），有点像洋葱的表皮。

② 插入语中常见的还有一些表示语气、态度、承上启下或转折的短语，这样的插入语有许多，典型的如：therefore (所以)，for example（例如），for instance（例如），in a conclusion

（作为结论），in short（简言之）等。但也有的插入语没有用逗号分开，要注意区分。

A robot rarely has two arms, **and even if it does**, they do not coordinate their movements the way our arms do. 机器人很少具有两个手臂，而且，即使它有两个手臂，这两个手臂也不像我们的两个手臂那样互相协调彼此的行动。

The area-delay trade-off involved here is, **in essence**, the problem of transistor size optimization. 这里所涉及的面积-延迟平衡（问题）本质上就是晶体管规模设计优化问题。

For many photographers who are going digital, **however,** the camera is only the first of a series of purchases. 但是对许多正在用数字相机的摄影师来说，相机只是要购买的一系列东西中的第一项。

15.4 Exercises

1. Put the Phrases into English

① 计算机键盘
② 连接成一个矩阵
③ 光标控制键
④ 帮助信息
⑤ 字母键
⑥ 自然的演变（发展）
⑦ 光电鼠标
⑧ 固化（在硬件中的）程序
⑨ 通信协议
⑩ 喷墨打印机

2. Put the Phrases into Chinese

① in business environments
② speedy data entry
③ function and control keys
④ the GUI-based Windows operating system
⑤ the signals from the sensors
⑥ an on-board digital signal processor (DSP)
⑦ photographic-quality output
⑧ laser printer
⑨ static electricity
⑩ the primary principle

3. Sentence Translation

① A computer keyboard is an array of switches, each of which sends the PC a unique signal when pressed.

② An important factor for keys is their force displacement curve, which shows how much force is needed to depress a key, and how this force varies during the key's downward travel.

③ As the use of computers in business environments increased, so did the need for speedy data entry.

④ Inside the mouse are a switch for each button, and a microcontroller, which interpret the signals from the sensors, and the switches, using its firmware program to translate them into packets of data, which are sent to the PC.

⑤ An inkjet printer is any printer that places extremely small droplets of ink onto paper to create an image.

⑥ The laser printer worked in a similar way to a photocopier, the difference being the light source.

⑦ With the powder pattern affixed, the drum rolls over a sheet of paper, which is moving along a belt below.

⑧ Because touchscreen systems are easy to use, overall training time for new employees can be reduced.

⑨ Self-service touchscreen terminals can be used to improve customer service at busy stores, fast service restaurants, transportation hubs, and more.

⑩ Initially, most commercial laser printers were limited to monochrome printing (black writing on white paper).

4. Translation

Here's a list of the core components that all video game systems have in common:
- User control interface
- CPU
- RAM
- Software kernel
- Storage medium for games
- Video output
- Audio output
- Power supply

The user control interface allows the player to interact with the video game. Without it, a video game would be a passive medium, like cable TV. Early game systems used paddles or joysticks, but most systems today use sophisticated controllers with a variety of buttons and special features. The two most common storage technologies used for video games today are CD and ROM-based cartridges. Current systems also offer some type of solid-state memory cards for storing saved games and personal information. Newer systems, like the PlayStation 2, have DVD drives.

15.5 课文参考译文

15.5.1 键盘

从本质上来说，键盘是一组开关，与一个微处理器相连接，微处理器管理每个开关的状态并在其状态发生改变时发出一个指定的响应（信号）。

按键一般有两种：机械式和橡皮膜式。机械式按键是简单的弹簧-负载（结构），当压下按键就接通电路，松手则断开电路。

橡皮膜式键盘由三层组成，第一层是有印刷线路的导体，第二层是上面有很多孔的隔离片，第三层是导体，带有很多凸出来的小块（键）。其中的橡皮膜给人有弹性感觉。当键被按下去时，键把两块导体压在一起使电路闭合，最上面塑料盖带有使键排列整齐的滑块。

键盘一个重要的参数是力-位移曲线，它显示了按下一个键需要多大的力和在键按下的过程中这个力是如何变化的。研究表明大多数人按键的力是 80～100 g，但玩游戏时按键的力可能大于 120 g。

按键按矩阵排列，行和列信号输给键盘自带的微处理器芯片，在键盘内部主板上的微处理器芯片用内部固化的程序处理这信号。如按下的键可能是第 3 行第 B（2）列，微处理器知道这是字母 A，就把与 A 相应的编码送给计算机。虽然各个键盘的行列定义可能不同，但这些"扫描码"在计算机的 BIOS 中有标准定义。

近年来因为（键盘制造商使他们的）键盘功能越来越强大，使键盘的固化程序越来越复杂。可编程的键盘（即其中有些键有可变换的多重功能）现在也很通用，这种键盘需要 8 KB 的 ROM 来存储固化程序。大部分可编程的功能是通过运行计算机中的驱动程序来执行的。

键盘布局自诞生以来变化极小，最多的变化只是简单地加一些键，以提供附加的功能。一个典型的键盘有四组键（如图 15.1 所示）：

- 打字键
- 数字键
- 功能键
- 控制键

打字键是键盘上的字母键，一般与（英文）打字机的布局相同。数字键是（根据需要）自然加出来的一部分。随着计算机在商业环境中应用的增加，需要提高数据输入的速度。因为大量的数据是数字，于是在键盘上加了一组 17 个键的数字键。这些键的布局与常用的加法器、计算器相同，使习惯于计算器等的会计可以方便地使用计算机。

在 1986 年，IBM 扩展了基本键盘，加上功能键和控制键。功能键是排在键盘的顶部第一行，可以用它调用应用程序或操作系统来完成特定的命令。控制键则用来控制屏幕和光标。在打字键和数字键之间的倒 T 形的四个光标键可使用户控制在屏幕上小范围地移动光标，一般控制键还有：Home 键、End 键、Insert 键、Delete 键等。

15.5.2 鼠标

在（20 世纪）80 年代早期，第一批计算机是只有传统输入设备——键盘，但到 80 年代后期，鼠标已成为基于 Windows 操作系统的计算机的基本配件。

最初用的最多的是（光电）机械鼠标，它的滚动球是钢球（有一定重量），外面覆盖一层橡皮（便于控制）。当滚动球旋转时带动两根滚子转动，一根滚子是确定 x 方向位移的，另一根是确定 y 方向位移的。

这两根滚子带动两个有刻缝的盘，每个盘在光发射-接收装置之间转动，每个光发射-接收装置有两个发光二极管和光敏传感器，当盘转动时，传感器接收到的光在闪动，表明有运动，而两个光敏传感器之间的偏移量则表明了运动的方向（如图 15.2 所示）。

在鼠标内部，相应每个按键都有一个开关，有一个微处理器分析来自传感器和开关的信号，用它的固化程序把信号转换成打包的数据送给计算机。串行口鼠标要 12 V 电压供电，按微软的协议，其打包的数据是三个字，报告 x, y 的方向和按键的情况。PS/2 鼠标和 USB 鼠标用 5 V 电压供电，按 IBM 开发的通信协议和专用接口（与计算机通信）。

1999 年引入了很多先进的鼠标设计，鼠标内部的滚动球和其他用于跟踪鼠标机械运动的运动部件不见了，取而代之的是微小的CMOS光电传感器（与数码相机中所用的一样）和主板上的数字信号处理器DSP。

光电鼠标的传感器和其他部件是这样工作的：

CMOS传感器把每幅图像送给DSP进行分析，DSP检测图像上的图案，求出与前一个图像相比图案移动了多少。根据一系列图像上的图案变化，DSP求出鼠标运动了多少，送出相应的坐标给计算机。计算机根据从鼠标中接收到的坐标在屏幕上移动（鼠标的）光标。这样的过程每秒钟要完成数百次，所以光标的移动非常平稳、流畅。

与机械鼠标相比，光电鼠标有很多好处：
- 没有运动部件，所以不易磨损，错误率低。
- 脏物不会进入鼠标，干扰传感器。
- 分辨率增加，即响应比较平稳。
- 不需要特别的垫板，例如鼠标垫。

15.5.3 喷墨打印机

过去喷墨打印机比激光打印机更受欢迎，因为它们可以打印彩色图片，所以是很普及的家用打印机。直到20世纪90年代后期，彩色激光打印机的价格降到了家用可以接受的水平，彩色喷墨打印机的优势开始减弱，但是，那时开发出的具有照片效果的图像输出使喷墨打印机在彩色王国中仍保持其优势。

喷墨打印机是在纸上喷出特别细小的墨点来产生一幅图像。如果你看到过喷墨打印机打出来的一页纸，你会发现：
- 墨点特别细小，（直径50～60 μm）比人的头发直径（70 μm）还要小。
- 墨点是非常精确定位的，其分辨率高达每平方英寸1 440×720个点。
- 不同颜色的墨点组合成具有照片一样质量的图像。

通常一台喷墨打印机主要有这样一些部件：
- 打印喷头——喷墨打印机的核心，打印喷头包括一组喷管（喷嘴）用来喷墨点。
- 墨盒——根据不同的厂商及型号，墨盒有各种形式。
- 喷头步进电动机——步进电动机移动打印喷头部分（喷头和墨盒）在纸上做横向往返运动。
- 纸输入盘或输入口——大部分喷墨打印机有一个放纸的输入盘，有些打印机省去标准的纸输入盘，而用一个输入口取代。
- 卷纸步进电动机——这个步进电动机拖动一个滚筒很精确地向前移动纸，以保证打印出一幅连续的图片。
- 控制电路——在打印机内部有一片复杂的小电路板，它控制打印的所有机械操作步骤以及对计算机送来的信息进行解码。

15.5.4 激光打印机

激光打印机是惠普（HP）公司在1984年在佳能开发的技术基础上造出的，它与复印机的工作方式相似，区别只是光源不同，复印机用白炽灯光扫描，而激光打印机当然是用激光作光源。

激光打印机的基本工作原理是静电效应，在干衣机中使衣服吸在一起或从雷云上打到地上的闪电都是静电引起的。静电效应是指在绝缘物体如汽球或你的身体上有电荷，因为异性电荷互相吸引，带有正负静电的物体会吸在一起。

激光打印机用这种现象作为一种"临时的胶水"，激光打印机的核心部件是感光器，一般是一个旋转的鼓（或圆柱形，中文常称硒鼓），这种鼓是用可以通过光子放电的高性能感光材料做的。

开始，流过电流的充电线环对鼓全部充正电荷，（有些打印机用充电滚筒代替充电线环，原理是一样的），当鼓转动时，打印机用一细激光束照在鼓的表面使一些点放电，这样，激光"画"出要打印的字符和图像成为一个电荷图案——一个静电图像，系统也可以用反电荷形式工作即正静电荷为图像，负电荷为背景。

（电子）图像建立后，打印机在鼓表面覆盖一层正电荷碳粉——一种细的黑色粉末。因为带正电荷，碳粉粘在鼓上的负电荷区域，但不与有正电荷的"背景"相吸，这有点像在汽水罐上用胶水写字，然后把瓶子放在面粉里滚一圈，面粉只粘在有胶水覆盖的地方，所以最后可看到用粉末所写的字。

粘好粉末图案后，鼓在一张纸上滚一圈（如图 15.3 所示），纸是下面的传送带送来的，当纸在鼓下面滚过之前先通过充电线环（或充电滚筒）充负电荷，这些电荷能量比静电图像的负电荷能量大，所以纸把（带正电荷的）颜色粉末带走，因为纸和鼓的运动速度相同，纸把图像原封不动地取走了。

最后，打印机让纸通过熔管（如图 15.4 所示）。当纸通过熔管（一对加热滚筒）时，松散的色粉熔解，与纸上的纤维熔在一起，熔管滚动着把纸送到出纸盒，打印完成了。当然熔管把纸也加热了，这就是刚从激光打印机或复印机出来的纸总是热的原因。

印过纸后，鼓表面通过放电灯，灯对整个感光器表面曝光，擦去静电图案，鼓表面再次通过充电线环充正电荷。

15.6 阅读材料参考译文

15.6.1 触摸屏

触摸屏是一种直观的计算机输入设备，通过用手指或一种笔简单地触摸显示屏，而不是敲键盘或用鼠标指点就可以输入信息。

触摸屏界面——使用户通过触摸计算机屏幕上的图标或连接标记可以浏览计算机，是最简单、直观、易学的计算机输入设备，所以很快就得到广泛应用，如：

公共信息系统：信息亭，旅游信息展示和其他电子显示系统可供很多缺乏计算机经验的人查询信息。尤其对新手来说，触摸屏友好的用户界面比其他输入设备都容易使用，因此触摸屏可以为最多的人提供信息。

旅社/电子收款机系统：时间就是金钱，尤其在一个快节奏的旅社或零售商店，因为触摸屏很容易使用，这些地方对新的雇员进行全面培训的时间就可以缩短，工作也可以做得更快。因为雇员可以简单地触摸屏幕来完成操作而不需要麻烦地用键盘输入命令了。

消费自服务系统：在当今世界快节奏的时代，排队等候也是应该加快的事情之一，在繁忙的商店，旅馆的快速服务、交通运输中心等处的自服务系统都可用触摸屏终端来改进服务（的方式和速度）。

控制/自动化系统：从工业过程控制到家庭（设备）的自动化工作，在很大的范围内触摸屏都是很有用的。通过在显示屏中集成了输入设备，可以节省工作空间。利用图形界面，操作者可以通过简单触摸屏幕实时监看和控制复杂的操作。

触摸屏由三个基本部件组成：

触摸屏传感器嵌板，它安放在显示器的上面，当触摸到某点时它产生相应的电压信号。

触摸屏控制器，用于处理传感器送来的信号，把触摸信号转换成相应的数据通过串行口或USB口送给计算机的处理器。

软件驱动程序，给计算机的操作系统提供一个界面，它把触摸事件数据转换成鼠标事件，本质上使传感器嵌板成为一个"仿"鼠标。

触摸屏有几种，这里谈一下红外触摸屏（如图15.5所示）。红外触摸屏是根据红外光束切断技术，用一个在四周围住屏幕的框架（产生-接收光束），而不是在屏幕表面加涂层的方法制作的。框架内含有印刷线路板，上面安放光电管，框架隐藏在一个红外光可以透过的斜面后面。

框架的一边含有（红外）光源或发光二极管，对面则有光传感器，其作用是在屏幕上形成一个光栅格，当任何物体触摸屏幕时，不可见光（红外光）被切断，导致光传感器接收的信号中断，根据是哪些光传感器没接收到（红外）光信号，就可以确定屏幕上被触摸的区域（坐标）。

红外触摸屏是固体状态技术，没有运动的部件，因此长时间使用传感器不会有磨损，再加上红外触摸屏不需要在屏幕上加涂层（这种涂层很容易破坏），所以不容易被破坏，且抗震性强。

15.6.2 彩色激光打印机

最初，大部分激光打印机是单色的（在白纸上打黑字）。但现在市场上有很多彩色激光打印机。

彩色打印机的工作方式与单色打印机基本相同，只是它们要完成四次完整的打印处理过程。每次处理一种颜色青（蓝）、品（红）、黄和黑色。通过以不同的比例组合这四种颜色，可以产生全部色彩。

彩色打印机有几种不同的处理方法，有些彩色打印机在一个转筒上有四种基色和显影单元（如图15.6所示），打印机把每种颜色碳粉放在相应的位置上打出一个静电图像，然后把这种颜色印到纸上，再接着处理下一种颜色。有些彩色打印机把四种颜色都放好再把图像印到纸上。一些更贵的打印机实际上对每种颜色有一个完整的打印单元——激光部分、鼓和碳粉系统，当纸通过不同的鼓头时，以一种流水线方式得到各种颜色的组合。

现在大部分激光打印机的分辨率为600或1 200点每英寸，彩色打印机的额定打印速度是每分钟3～5页，单色打印机是每分钟12～14页。1998年秋，Lexmark公司生产出每分钟可打印12页的发光二极管的彩色打印机，在这个发展的关键区域起到领先作用。他们采用同时处理四种颜色、一次打印处理方法，声称其速度已达到单色打印机的水平。

除了速度问题，彩色打印机的主要优点是输出稳定，这是由碳粉的（化学）惰性决定的，碳粉是熔在纸的表面，而喷墨打印机的墨水是被纸吸收，因此彩色激光打印机可以在各种纸上打印，颜色不会互相染污，不会褪色，而这些是喷墨打印机难以解决的问题。再加上通过控制加热过程中的热量和压力，用户可以控制输出的纸表面产生从粗糙到光滑（有光泽）的纹理效果。

15.6.3 电子电路仿真

电子电路仿真是用数学模型去模仿一个实际电子器件或电路的性能。在构建实际电路前先仿真这个电路的性能，可以预先了解电子电路设计的性能，提高设计效率。实际上，对集成电路来说，加工（光掩膜）是比较贵的，面包板则不实用，而且要了解内部信号的性能十分困难。所以几乎所有的集成电路设计都要先仿真。众所周知，最常用的模拟仿真器（软件）是 SPICE，而数字仿真器则是那些基于 Verilog 和 VHDL 的软件。

有些电子仿真器集成了原理图编辑器、仿真器，可在屏幕上显示波形，使得假设仿真（如果……，则……）变得容易和直接。通常这些仿真器还含有很多模块和器件库。这些模块包括集成电路专用的晶体管模块，通用器件如电阻、电容、电感和变压器、用户定义模块。印制板（PCB）设计还需要专用模块，如画导线的传输线模块和驱动及接收电子的 IBIS 模块。

虽然有严格的模拟电子电路仿真器，但大众化的仿真器常常包括模拟和数字仿真功能，称为混合模型仿真器。这意味着任何仿真可以含有模拟器件、事件驱动器件（数字或采样数据）或两者的组合。可以对一个集成电路原理图进行完整的混合信号分析。在混合模块仿真器中给出所有的数字模块的传播时间和上升/下降延迟等准确描述（说明）。

这里以一个混合仿真器——电子工作台为例。电子工作台（EWB）是一个设计工具，它提供所有的元件和必要的仪器，用以进行电子线路的设计。图 15.7 是 EWB 软件的界面。它可以完成模拟信号和数字信号的混合电路仿真和波形分析。你可以用它设计电路并用不同的仿真仪器和分析选项进行分析。软件是完成集成的，有很好的交互性，因此你可以很快地修改电路，进行快速和反复的假设分析。

EWB 仿真器，就像其他通用仿真器一样，有 4 个主要的步骤：输入、设置、分析和输出。
- 在输入阶段，建立一个电路，指定参数值，然后选取一种分析方法，仿真器读入电路的信息。
- 在建立阶段，仿真器构造并检查对电路作完整描述的一组数据结构。
- 在分析阶段，进行在输入阶段指定的电路分析，这个阶段占用电脑最多的执行时间，实际上是电路仿真的核心。分析阶段对指定的分析建立方程并对电路方程组求解，给出全部数据，这些数据可以直接输出或留在后期处理。
- 在输出阶段，就可以看到仿真结果了。可以在仪器如示波器上看到结果，也可以用图片显示结果，只要你在分析菜单上选取一种分析或者选择分析/显示图片。

Unit 16 Multimedia Technology

 Pre-reading

Read the following passage, paying attention to the question.
1) What is Multimedia?
2) What does multimedia include?
3) What can we do with multimedia?

16.1 Text

16.1.1 What is Multimedia

Some elements of media are: animation, sound, graphics, text, video, photography.

Multimedia involves the combination of two or more media types to effectively create a sequence of events that will communicate an idea, usually with both sound and visual support. Typically, multimedia productions are developed and controlled by computer.

Multimedia isn't new, and the term has been used for decades to describe slide presentation accompanied by audio tape (slide/tape). The combination of slide and narration has been both a popular and successful form of business presentation.

In the 1970s the slide show format was introduced to the computer, this technology allowed the computer to control numerous projectors, coordinating them in a manner that produced fast-paced dissolves and effects. Taped soundtracks would contain cues that triggered the slide projectors to do what it was programmed to do.

In the 1980s PCs were designed to "cut" a graphic element and "paste" it into another document. Since then software and hardware developers have been scrambling to integrate various forms of media into the personal computer.

16.1.2 Multimedia Assets

Effective multimedia applications depend on the most effective use of various materials - referred to as assets or resources. There are various multimedia resources:

1. Text

There are 3 major advantages generally associated with screen-based text compared with paper-based text: the ability to spontaneously update the screen, the reactive capability, and the ability to incorporate special effects.

2. Audio

Interactive audio can add a particular dimension of reality to multimedia systems. Until recently, audio-based subject domains - such as music, linguistics, languages etc - have been virtually ignored. Also, the efficiency of interaction between the human and the machine could have been greater had sonic interaction and audio enhancements been available sooner.

3. Pictures

When using images, the designer must decide on the most economical and efficient method to generate them. At the planning stage of the application the designer must decide the most appropriate combination of technology and software to be able to reproduce and create new images.

4. Video Images

Portable video recorders have made it relatively easy to capture real-time video images. Video can be incorporated into multimedia applications using two different processes. The first involves using a video source connected to the computer via a controller card. This technology has been referred to as interactive video. A more integrated process converts video from analogue into digital format that can be manipulated by the desktop computer.

16.1.3 Multimedia Applications

There are various ways to group multimedia titles. They can be classified by *market* (such as home, business, government and school); by *user* (such as child, adult, teacher and student); or by *category* (such as education, entertainment, and reference). Here we group them into seven application areas and to look at each in a more detail.

1. Reference

Encyclopedias, census data, yellow pages, atlases and street directories are examples of CD reference titles. In many cases they are electronic versions of reference books. The challenge for the developer is to make it easy for the user to find the desired information and to effectively use other multimedia elements such as sound, video and animation.

2. Education

The goal of the educator is to facilitate learning - to help the student gain a body of knowledge, acquire specific skills and function successfully in society. But one of the greatest challenges to educators is the diversity of students; especially in the different ways they learn. Some students learn better through association, others by experimentation, some are more visually oriented, others are more auditory.

Multimedia has the ability to accommodate different learning styles and can present material in a non-linear manner. It is motivating, it can be highly interactive, and it can provide feedback and evaluate skills.

3. Training

Every company has a need to train its employees on a wide range of subjects from personnel policy to equipment maintenance.

With multimedia the trainee can perform a simulated job function in order to develop an advanced level without having touched the actual unit. The integration of audio and video allows this training technology to be a highly effective medium in areas such as flight and driving simulators. Similarly, NASA uses multimedia extensively for flight control training for astronauts.

4. Entertainment

Drawing the line between education and entertainment in multimedia can be almost impossible, hence the term 'edutainment'. Multimedia *can* make learning entertaining.

But multimedia also has a purely *entertainment side*. Anything that's possible in sound and images is possible on a multimedia CD. For an example, *AIATSIS* is an encyclopedia of the Australian Aborigine containing over 2 000 entries, 1 000 photos, 230 sound clips and 50 videos. It covers subjects ranging from art to health, from technology to law, which is integrated in a CD.

5. Businesses

As businesses have the need to communicate with the outside world, multimedia processes offer a wide variety of options for business presentations, marketing and sales. Multimedia can be used at trade shows or to produce electronic catalogues. The marketing of new products can be greatly enhanced by using multimedia; these products can be marketed in a manner that will provide more detailed and stimulating information than printed media.

6. Presentation

Thousands of multimedia presentations are made in the business world every day. Company CEOs give their annual report to a meeting of stockholders. Sales representatives pitch their product line to a group of potential customers. A conference speaker tells an audience about industry trends. From an electronic slide show to an interactive video display multimedia can enhance a presentation.

Multimedia provides the presenter with the tools to attract and focus the audience's attention, reinforce key concepts and enliven the presentation.

7. Interactive Game

Multimedia means interaction, and too many interactive entertainment means games. Game developers were the pioneers in the use of multimedia and still provide the most innovative and interactive applications of multimedia.

In order to attract, engage, captivate and challenge the user multimedia provides the fast action, vivid colours, 3D animations and elaborate sound effects that are essential to entertainment. It can also provide the rewards, recognition and sense of accomplishment that are often part of entertainment titles.

Many games have moved from the physical (hand/eye coordination) to the mental (solving the mystery, overcoming evil, outwitting the opponent).

On the other hand, hobbies and sports are examples of multimedia titles that provide the user with a vicarious experience such as being able to play the best golf courses in the world or simulate flying over 3D cityscapes.

Technical Words and Phrases

aborigine	[æbə'rɪdʒɪnɪ]	n. 土著，土著居民，Aborigine 澳大利亚土著居民
animation	[ænɪ'meɪʃ(ə)n]	n. 动画，活跃；生气勃勃；兴奋
assets	['æsets]	n. 资产
cityscape	['sɪtɪskeɪp]	n. 都市风景
coordinate	[kəu'ɔːdɪneɪt]	vt. 调整，整理　n. 同等者，坐标（用复数）　adj. 同等的，并列的
dissolve	[dɪ'zɔlv]	n. 渐渐消隐，溶化　vt. 叠化，叠化画面
document	['dɔkjuːment]	n. 公文，文件，文档，档案，文献　v. 证明
effect	[ɪ'fekt]	n. 结果，效果，作用，影响，（在视听方面给人留下的）印象
encyclopedias	[ensaɪkləu'piːdɪə]	adj. 如百科辞典的，百科全书式的
incorporate	[ɪn'kɔːpəreɪt]	adj. 合并的，结社的，一体化的　vt. vi. 合并，使组成公司，具体表现
maintenance	['meɪntɪnəns]	n. 维护，保持，生活费用，扶养
manipulate	[mə'nɪpjuleɪt]	vt.（熟练地）操作，使用（机器等）　vt.（熟练地）操作，巧妙地处理
multimedia	[mʌltɪ'miːdɪə]	adj. 多媒体：在屏幕上输出的文本、图形、视频（照片）和音频（声音）等的合成
narration	[næ'reɪʃ(ə)n]	n. 讲述，叙述，叙述故事
outwit	[aut'wɪt]	vt. 瞒骗，以智取胜
presentation	[prezən'teɪʃ(ə)n]	n. 介绍，陈述，赠送，表达
projector	[prə'dʒektə(r)]	n. 电影放映机；幻灯机；投影
reference	['ref(ə)rəns]	n. 提及，涉及，参考，证明书（人），介绍信（人）
slide	[slaɪd]	v.（使）滑动，（使）滑行　n. 滑，滑动，幻灯片
soundtrack	['saundtræk]	n. 音轨
spontaneously	[spɔn'teɪnɪəslɪ]	adv. 自然地，本能地　adv. 自发地，自生地
style	[staɪl]	n. 风格，时尚，文体，风度，类型，字体　vt. 设计
trigger	['trɪgə(r)]	vt. 引发，引起，触发　n. 板机

16.2 Reading Materials

The hardware of multimedia is including graphics Cards, CRT monitors, panel displays, sound cards, CD and CD player, digital video disc (DVD) and DVD player.

16.2.1 DVD

A DVD is very similar to a CD, but it has a much larger data capacity. A standard DVD holds about seven times more data than a CD does. This huge capacity means that a DVD has enough room to store a full-length, MPEG-2-encoded movie, as well as a lot of other information. Here are the typical contents of a DVD movie:

- Up to 133 minutes of high-resolution video, with 720 dots of horizontal resolution (The video compression ratio is typically 40:1 using MPEG-2 compression.)
- Soundtrack presented in up to eight languages using 5.1 channel Dolby digital surround sound
- Subtitles in up to 32 languages

DVD can also be used to store almost eight hours of CD-quality music per side.

DVDs are of the same diameter and thickness as CDs, and they are made using some of the same materials and manufacturing methods. Like a CD, the data on a DVD is encoded in the form of small pits and bumps in the track of the disc.

A DVD is composed of several layers of plastic, totaling about 1.2 millimeters thick. For single-sided discs, the label is silk-screened onto the nonreadable side. Double-sided discs are printed only on the nonreadable area near the hole in the middle. Each writable layer of a DVD has a spiral track of data. On single-layer DVDs, the track always circles from the inside of the disc to the outside. That the spiral track starts at the center means that a single-layer DVD can be smaller than 12 centimeters if desired.

You may wonder how incredibly tiny the data track is - just 740 nanometers separate one track from the next (a nanometer is a billionth of a meter), and the elongated bumps that make up the track are each 320 nanometers wide, a minimum of 400 nanometers long and 120 nanometers high.

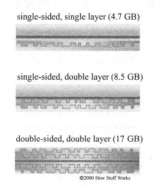

Fig 16.1 DVD formats

To increase the storage capacity even more, a DVD can have up to four layers, two on each side (Fig 16.1). The laser that reads the disc can actually focus on the second layer through the first layer. Even though its storage capacity is huge, the uncompressed video data of a full-length movie would never fit on a DVD. In order to fit a movie on a DVD, you need video compression. A group called the Moving Picture Experts Group (MPEG) establishes the standards for compressing moving pictures.

A DVD player is very similar to a CD player. It has a laser assembly that shines the laser beam onto the surface of the disc to read the pattern of bumps. The DVD player decodes the MPEG-2 encoded movie, turning it into a standard composite video signal. The player also decodes the audio stream and sends it to a Dolby decoder, where it is amplified and sent to the speakers.

The DVD player has the job of finding and reading the data stored as bumps on the DVD. Considering how small the bumps are, the DVD player has to be an exceptionally precise piece of equipment. The drive consists of three fundamental components:

A drive motor to spin the disc - the drive motor is precisely controlled to rotate between 200 and 500 rpm, depending on which track is being read.

A laser and a lens system to focus in on the bumps and read them - the light from this laser has a smaller wavelength (640 nanometers) than the light from the laser in a CD player (780 nanometers), which allows the DVD laser to focus on the smaller DVD pits.

A tracking mechanism that can move the laser assembly so the laser beam can follow the spiral track - the tracking system has to be able to move the laser at micron resolutions.

Inside the DVD player, there is a bit of computer technology involved in forming the data into understandable data blocks, and sending them either to the DAC, in the case of audio or video data, or directly to another component in digital format, in the case of digital video or audio.

16.2.2 News: Vivid Animations Help Students with Science

Students learning science in the classroom can now give their textbooks a break.

One of the best places for hands-on learning is in science class, but small budget experiments and textbooks don't always get the point across.

"It's too much text based, too much reading about difficult concepts and for a lot of students, it's very hard for them to see what those words mean," said Linda Cauley, Director of Shenandoah Valley Governor's School. Now, a new virtual reality website helps students get a better grasp on basic science concepts.

"I'm a very visual learner - it's a lot easier for me to memorize say a picture or a graph as opposed to a paragraph of letters," said Ben Holman, who is a high school student.

"We were learning in the textbook about it, but then we went to the computers and showed a different way to look at it," said Katie Blackard, another high school student. The virtual lab, developed by engineers and students at the University of Virginia, guides students through 50 experiments (Fig 16.2), along with text and vivid animations that explain how things work - like semiconductors and generators.

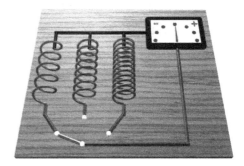

Fig 16.2 virtual experiment

"The thing about virtual reality is I can show things that are invisible. One of the things that we depict frequently is electrical field. You can't see those. I can make them visible with virtual reality," said John Bean, an engineer at the University of Virginia.

The site is geared to high school and college freshmen - providing online study help for homework assignments - for a better, visual understanding of science.

"Seeing it on the computer screen makes it a better picture in your head, a better way to understand it," said Katie Blackard.

16.2.3 about Graphics Cards

Video or graphics circuitry, usually fitted to a card but sometimes found on the motherboard itself, is responsible for creating the picture displayed by a monitor. On early text-based PCs this was a fairly mundane task.

As the importance of multimedia and then 3D graphics has increased, the role of the graphics card has become ever more important and it has evolved into a highly efficient processing engine which can really be viewed as a highly specialized co-processor. By the late 1990s the rate of development in the graphics chip arena had reached levels unsurpassed in any other area of PC technology. One of the consequences of this has been the consolidation of major chip vendors and graphics card manufacturers.

The modern PC graphics card consists of four main components:
- the graphics processor
- the video memory
- the random access memory digital-to-analogue converter (RAMDAC)
- the driver software

There are two main parameters about graphics card.

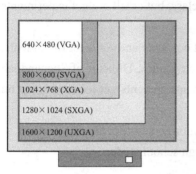

Fig 16.3   displayable pixels in a display

1. Resolution

Resolution is a term often used interchangeably with addressability, but it more properly refers to the sharpness, or detail, of the visual image. It is primarily a function of the monitor and is determined by the beam size and dot pitch (sometimes referred to as "line pitch"). A complete screen image is composed of thousands of pixels and the screen's resolution - specified in terms of a row by column figure - is the maximum number of displayable pixels (Fig 16.3). The higher the resolution, the more pixels that can be displayed and therefore the more information the screen can display at any given time.

2. Colour Depth

Each pixel of a screen image is displayed using a combination of three different colour signals: red, green and blue. The precise appearance of each pixel is controlled by the intensity of these three beams of light and the amount of information that is stored about a pixel and determines its colour depth. The more bits that are used per pixel ("bit depth"), the finer the colour detail of the image.

For a display to fool the eye into seeing full colour, 256 shades of red, green and blue are required; that is 8 bits for each of the three primary colours, hence 24 bits in total. However, some graphics cards actually require 32 bits for each pixel to display true colour, due to the way in which they use the video memory.

16.2.4   DirectX

Graphics technology is a particularly fast-developing area of the PC industry, with new chipsets, new revisions of chipsets and even entirely new technologies appearing at an alarming rate. This presents a problem for applications wishing to take advantage of the latest 3D hardware, as it's absolutely impossible for any application developer to write native code for every graphics processor.

The solution is an API (application programming interface). APIs act as an intermediary between application software and the hardware on which it runs. The software vendor writes code that outputs its graphics data to the API driver via standardized commands, rather than directly to the hardware. The driver, written by the manufacturer of the hardware, then translates this standard code to the native format understood by a particular model of hardware.

First introduced in 1995, DirectX is an integrated set of programming tools designed to help developers create a whole range of multimedia applications for the Windows platform. It covers almost all aspects of multimedia content and by the time of DirectX 7.0 - its sixth major release introduced in 1999 - comprised the following main components:

Direct3D - used real-time 3D graphics.

DirectDraw - used accelerated 2D graphics.

DirectSound - used for audio playback.

DirectPlay - used for network connectivity (especially for Internet multi-player gaming).

DirectInput- used for joysticks and other related devices.

DirectMusic - used for message-based musical data.

Now the DirectX 9.0 is widely used and has more versions for various languages.

16.3 Knowledge on Writing a Research Paper （科技论文写作知识）

A research paper is a form of written academic communication that can be used to give useful information and to share academic ideas with others. Most of the research papers are written for publication in journals or conference proceedings in one's field. Publication is one of the fastest ways for propagating ideas and for professional recognition and advancement. If you have clear idea about the features and styles get your paper published in the target journal or accepted by an international conference.

一篇研究论文是指用写作方式进行学术交流的一种形式，它用来提供有用信息和与其他人共享学术思想。大部分学术论文都是为某一学术领域的出版物（杂志）或某一领域的学术会议而写的。出版物是传播学术思想、受到学术界公认和（倡导）学术进步的最快途径之一。如果你在写论文时对论文的格式和特征很了解的话，你的论文就能在学术杂志上发表或被一个国际学术会议认可（允许你在会上宣读你的论文）。

Features of Academic Papers（学术论文的特征）

The first of the features of an academic paper is the content. It is no more and no less than an objective and accurate account of a piece of research you did, either in the humanities, social sciences, natural sciences or applied sciences. It should not be designed to teach or to provide general background.

The second feature is the style of writing for this purpose. Your paper should contain three ingredients: precise logical structure, clear and concise language, and the specific style demanded by the journal to which it will be submitted. From the instructions on manuscripts in the "Appendixes A to E" you may get a brief idea of different styles required by different journals.

The third, which is indeed a part of the second, is the system of documenting the sources used in writing the article. At every step in the process of writing, you must take into account the ideas, facts, and opinions you have gained from sources you have consulted.

One of the most convenient features of academic articles is that they are divided into clearly delineated sections. This is helpful because you only have to concentrate on one section at a time. You can thus visualize more or less completely the whole paper while you are working on any part of it. Though papers of the humanities and social sciences do not always have the clearly divided sections, they share some of the common requirements with the scientific papers.

学术论文的第一个特征是内容，每篇论文只要叙述一个观点，准确描述你所进行的研究中的一个部分，无论是人文科学、社会科学、自然科学或应用科学。不要把论文写成像教材或给出一般的背景（泛泛而谈）。

学术论文的第二个特征是为写作目的所采用的写作格式。论文应包含三个部分，合理的逻辑结构、清晰精练的语言和按准备投稿的杂志所要求的特定的格式。不同的杂志有不同的"投稿须知"，你可以通过它来了解各种杂志对格式的要求。

学术论文的第三个特征实际上是第二个特征的一部分，是在写文章中所用的资料系统，在写作过程的每一步中，你都要考虑你已从查阅、参考的资料中所获取的思想，（实验，试验）事实和观点。

学术论文最方便的特征之一是学术论文清晰地分成几个部分。这一点很有用，因为每次你只需集中考虑一个部分。当你撰写任一部分时，你就可以看到整篇论文在逐步形成。虽然人文学和社会科学（的学术论文）的几个部分之间并不总是分得很清楚，但它们都和科学论文一样共同遵守某些普遍要求。

Divisions of Academic Papers

For the average scientific paper the following suggested outline of the divisions of a paper is normally acceptable to, and demanded by, the editors of journals or compilers of conference proceedings:

一般学术会议或杂志的编辑要求学术论文按如下的大纲分成几个部分：

① Title of the Paper (subtitle if necessary)：标题，副标题。

② By lines：标题下写作者的名字和地址。

Name(s) of author(s)

Affiliation(a) of author(s): present and /or permanent address

③ Abstract：摘要。

The purpose and scope of the paper   论文涉及的范围及写作目的

The method of study or experiment   实验研究的方法

A very brief summary of the results, conclusion, and/or recommendations   结果、结论或建议的一个简短的总结介绍

④ Introduction：引言。

A statement of the exact nature of the problem   问题本质的介绍

The background of previous work on this problem done either by the author or others of different approaches   关于这个问题其他人或作者做的前期工作的情况，或采用其他方法作的情况

The purpose of this paper   这篇论文的目的

The method by which the problem will be attacked   处理问题的方法

The primary findings, conclusions and significance of this work   主要的发现，结论和这项工作的意义

A statement of the organization of the material in the paper　论文中材料的组织方法介绍

⑤ Body of Paper：论文主体。

The organization of this main part of the paper is left to the discretion of the author. The information should be presented in some logical sequence, the major points emphasized with suitable illustrations, and the less important ideas subordinated in some appropriate way. This portion of the paper should be styled for the specialist and should not be designed to teach or to provide background for the general reader.

论文的主要部分的组织方法是作者自己确定的，文中所提供的信息应按逻辑顺序，主要观点要强调，并加适当的说明，不重要的观点则用适当的附属的方法表示（注意突出重点）。这部分的内容是给专家（同行）看的，不要把论文写成像教材或给出一般的背景（泛泛而谈）。

⑥ Conclusion：结论。

Summary and evaluation of the results　小结并评价（实验）结果

Significance and advantages over previous work　这些工作的意义和优点

Gaps and limitations in the work　工作中的局限性和存在的问题

Directions for future work and applications　对未来工作和应用的描述

⑦ Acknowledgments：致谢，对论文（研究）有帮助的人表示感谢

⑧ References：参考文献。

16.4　Exercises

1. Put the phrases into English

① 幻灯片
② 商业介绍
③ 百科全书
④ 找到想要的信息
⑤ 提供反馈（信息）
⑥ 模拟驾驶器
⑦ 教育和娱乐
⑧ 销售代表
⑨ 逼真的色彩
⑩ 三维动画

2. Put the phrases into Chinese

① create a sequence of events
② multimedia assets
③ the most economical and efficient method
④ the most appropriate combination of technology and software
⑤ portable video recorders
⑥ digital video
⑦ learn better through association
⑧ have a purely entertainment side
⑨ the pioneer in the use of multimedia
⑩ the integration of audio and video

3. Sentence Translation

① Multimedia involves the combination of two or more media types to effectively create a sequence of events that will communicate an idea, usually with both sound and visual support.

② Since then software and hardware developers have been scrambling to integrate various forms of media into the personal computer.

③ The efficiency of interaction between the human and the machine could have been greater had sonic interaction and audio enhancements been available sooner.

④ This approach (called digital video) allows all video operations - including editing and special effects - to be carried out on the computer.

⑤ DVDs are of the same diameter and thickness as CDs, and they are made using some of the same materials and manufacturing methods.

⑥ You may wonder how incredibly tiny the data track is - just 740 nanometers separate one track from the next (a nanometer is a billionth of a meter).

⑦ To increase the storage capacity even more, a DVD can have up to four layers, two on each side.

⑧ Video or graphics circuitry, usually fitted to a card but sometimes found on the motherboard itself, is responsible for creating the picture displayed by a monitor.

⑨ A drive motor to spin the disc - the drive motor is precisely controlled to rotate between 200 and 500 rpm, depending on which track is being read.

⑩ Graphics technology is a particularly fast-developing area of the PC industry, with new chipsets, new revisions of chipsets and even entirely new technologies appearing at an alarming rate.

4. Translation

What is the difference between two- and three-pronged plugs?

Let's start with what the holes in an outlet do. When you look at a normal 120-volt outlet in the United States (Fig16.4 (a)), 220 V outlet in china (Fig 16.4(b)), there are two vertical slots and then a round hole centered below them. The left slot is slightly larger than the right. The left slot is called "neutral," the right slot is called "hot" and the hole below them is called "ground." The prongs on a plug fit into these slots in the outlet.

(a) used in U.S.A

(b) use in China

Fig 16.4 outlet

If you look around your house, what you will find is that just about every appliance with a metal case has a three-prong outlet. This may also include some things, like your computer, that have a metal-encased power supply inside even if the device itself comes in a plastic case. The idea behind grounding is to protect the people who use metal-encased appliances from electric shock. The casing is connected directly to the ground prong.

Let's say that a wire comes loose inside an ungrounded metal case, and the loose wire touches the metal case. If the loose wire is hot, then the metal case is now hot, and anyone who touches it will get a potentially fatal shock. With the case grounded, the electricity from the hot wire flows straight to ground, and this trips the fuse in the fuse box. Now the appliance won't work, but it won't kill you either.

What happens if you cut off the ground prong or use a cheater plug so you can plug a three-prong appliance into a two-prong outlet? Nothing really - the appliance will still operate. What you have done, however, is disable an important safety feature that protects you from electric shock if a wire comes loose.

16.5　课文参考译文

16.5.1　什么是多媒体

媒体元素是指：动画、声音、图像、文本、视频、照片等。

多媒体是指结合两种或多种媒体（元素），通常有声音和视频支持，达到生动地表达一种思想（概念）的作品。一般多媒体产品用计算机开发和控制。

多媒体并不是新词，这个词已用了几十年，指放幻灯片时伴随声音解释。在商业介绍中幻灯和叙述的结合已经十分普及和成功。

在（20世纪）70年代，幻灯播放方式被引入到计算机中，计算机可控制多个投影仪，计算机技术控制、协调这些投影仪产生一种快节奏的切换方式和生动的效果。磁带声道中含有信号，这些信号触发投影仪按事先所设置的方法放映。

到（20世纪）80年代，计算机已设计成可以切割图像元素和把图像元素"粘"到另一个文件中，从此以后，软件和硬件开发商们开始争着把各种媒体形式集成到个人计算机中。

16.5.2　多媒体资源

要获得生动有效的多媒体效果，就必须充分运用称为（多媒体的）资源的各种材料。多媒体资源有：

1．文本

与纸质文本相比较，在屏幕上文本有三个主要的优点：能随时刷新屏幕、能响应（读者的需求）和可以产生特殊的伴随效果。

2．声音

交互的声音可以给多媒体系统增加一个特别真实的元素。以前，基于声音的领域如音乐、语言等实际上被忽略了，如果能更快地实现声音的交互和音频（质量）的改进，则人和计算机之间的交流将变得更加生动。

3．图片

要使用图片，设计者必须选择一种最经济有效的方法去制作图片。在应用（多媒体）的计划阶段，设计者必须选择技术和软件最适当的组合，以再现和产生新的图像。

4．视频

手提视频录像机可以很方便地拍摄实时视频图像。通过两个不同的方法可以把视频与多媒体应用相结合。第一个方法是通过控制卡把视频源（如录像机等）连接在计算机上，这种技术称为交互视频。更完整的方法是把视频从模拟（数据）转换成可以用计算机处理数字的格式（数据）。

16.5.3 多媒体的应用

多媒体有各种分类方法，可以按市场分（如家用、商业、政府和学校），（也可以）按用户分（孩子、成年人、教师、学生）或按用途分（如教育、娱乐、参考资料等）。这里我们按用途把它分成 7 个应用领域，并一一详细阐述。

1. 参考资料

CD 参考资料有很多，如百科全书、调查资料、黄页（电话查号簿）、地图集和街道目录等。通常是参考（资料）书的电子版本。开发者要做的是使用户可以很方便地查找想要的信息和有效地利用其他的媒体元素如声音、视频和动画（使它们变得更加生动）。

2. 教育

教育的目标是促进学习——帮助学生获得整体的知识，掌握某种技能和成功地立足于社会。对教育者来说其中一个难题是学生尤其是他们的学习方法是各种各样的。有些学生通过联想学得比较好，有些学生则通过实验学得比较好，有些喜欢多看，有些喜欢多听。

多媒体可以适应不同的学习模式，可以用非线性化的方式（例如超链接）显示（学习）材料，它是激发式的，可以与学生交流，还可以提供（信息）反馈和评价（学生）技能。

3. 培训

每个公司需要从人员政策到设备维护等很多方面培训它的雇员。

受培训者可以用多媒体完成仿真的工作任务，在不接触实际设备的情况下，进行高级培训。例如在飞行和驾驶领域中进行技术培训时，声音和视频的结合可使这种培训非常生动。美国宇航局就用这种多媒体技术对宇航员进行大量的飞行控制训练。

4. 娱乐

很难划分多媒体中教育与娱乐的界线，因此就有了新词"寓教于乐"，多媒体可以使学习变得有趣。

但多媒体也有纯娱乐的一面，声音和图像所具有的任何效果在一张多媒体 CD 上都可以实现。例如 *AIATSIS* 是一本澳大利亚的百科全书，含有 2 000 多条目、1 000 张照片、230 个声音片段和 50 段视频录像，其主题覆盖了艺术、健康、技术和法律知识。所有这一切都收录在一张 CD 上。

5. 商业

商业有与外部世界交换信息的需要，多媒体为商品介绍、市场和销售推广提供了很大的选择范围，多媒体可用在传统的展示方面（制作展板等），或提供电子（销售）目录。利用多媒体可以增强新产品的市场推广，多媒体可以比印刷品提供更详细和更吸引人的产品信息。

6. 演讲（展示）

每天在商界中举行着成千上万个多媒体演讲展示会。公司的执行总裁在股东会议上作年度报告，销售代表向有（购买）潜力的顾客兜售产品。一个会议的演讲人在给听众介绍工业（发展）趋势。从电子幻灯放映到可以互动的视频展示，多媒体使演讲多姿多彩。

多媒体给演讲人提供了可以吸引听众视线、增强关键概念提示和使演讲变得生动的工具。

7. 互动游戏

多媒体意味着互动，许多互动娱乐则意味着游戏。在利用多媒体技术方面游戏开发商总是走在前面，现在还在不断提供多媒体的最新的和互动的应用技术。

为了吸引玩家、迷住玩家和挑战玩家，多媒体技术提供了娱乐所必需的最快的动作、逼真的色彩、三维动画和精美的声音效果，它还提供了娱乐常有的奖励、荣誉和成就感。

许多游戏从动作（手、眼配合）到脑力（解谜、战胜恶魔、与对手斗智），在各个方面与玩家斗智斗勇。

另一方面，多媒体也可以为业余爱好和模拟运动设计游戏，如让玩家作各种体验，例如在世界上最好的高尔夫球场打高尔夫球或在三维的城市高空作模拟飞行。

16.6 阅读材料参考译文

多媒体的硬件包括图像卡、CRT 显示器、平板显示器、声卡、CD 和 CD 播放器、数字视频盘（DVD）和 DVD 播放器

16.6.1 DVD

DVD 与 CD 很相似，只是数据（存储）容量更大。一张标准的 DVD 存储的数据是 CD 的 7 倍，这么大的容量意味着一张 DVD 可以存储一部完整的 MEPG-2 格式编码的电影，包括一些其他的信息。一般存储一部 DVD 电影要包括以下内容：

- 大约 133 分钟以上的 720 线的水平分辨率的高清晰视频（视频压缩比采用 MPEG-2 格式，一般是 40:1）。
- 采用道尔比（道尔比降噪系统—商标名）5.1 频道数字环绕立体声，有 8 种语言声道。
- 多达 32 种语言的对白字幕。

DVD 每面可以存储约 8 小时 CD 质量的音乐。

DVD 的直径、厚度与 CD 相同，它们用一些与制作 CD 相同的材料和制作方法进行制作，与 CD 一样，在 DVD 的数据是以凹凸的轨道来编码的（如图 16.1 所示）。

DVD 是由几层塑料制成的，约 1.2 mm 厚，单面盘可以在不能读的一面作标记，而双面盘只在中心孔附近不能读的地方作标记。每个 DVD 的可写层都有数据的螺旋轨道，在单层 DVD 盘上，轨道是从内到外的圆。螺旋轨道从中心开始，所以如果需要的话，单层 DVD 直径可以做得小于 12 cm。

令人惊奇的是这些数据轨道细得无法使人相信，两个轨道之间只有 740 nm（1 nm 是十亿分之一米）。每条用于制作轨道的凸道宽 320 nm，最小 400 nm 长，120 nm 高。

为了存储更多数据，DVD 已发展到有四个存储层，每面两层（如图 16.1 所示）。读盘的激光实际上可以透过第一层聚焦在第二层上。虽然 DVD 的存储量已非常大，还是存储不下一部没有压缩的电影。为了在 DVD 上存储一部电影，需要视频压缩，因此，一个称为动态画面专家小组（MEPG）的组织建立了动态画面的压缩标准。

DVD 播放器与 CD 播放器很相似，其中有一个激光器发出激光照在 DVD 盘的表面，读出凹凸图案（数据）。DVD 播放器对 MEPG-2 编码的电影数据解码，把它转化成一个标准的复合视频信号。DVD 播放器还对声音（流）解码，把它送到道尔比解码器，在那里进行放大并送给喇叭。

DVD 播放器有读出凸出轨道上存储的数据的功能，想到这些凸起的轨道是多么细，因此

DVD 播放器必须是一个特别精密的设备，DVD 驱动由三个基本部分组成：

旋转 DVD 盘的驱动电动机——根据所读的轨道，驱动电动机精确控制在每分钟 200～500 转。

聚焦及读取轨道上数据的激光和透镜系统——发出的激光波长（640 nm）比 CD 播放器的激光波长（780 nm）短，使 DVD 激光可以聚焦在更小的 DVD 凹凸块上。

轨道机械部分——用以移动激光单元让激光束可以沿着螺旋形的轨道移动，这个轨道系统必须以微米级分辨率来移动激光（定位很精确）。

在 DVD 播放器内部，涉及的计算机技术把数据转换成可理解的数据块然后送到 DAC（数模转换器）转换成音频或视频数据，或者直接送到数字视频或数字音频这类数字设备中。

16.6.2 新闻：学生通过生动的仿真（实验）学科学

在教室里学习科学知识的学生现在可以放下书本了。

最好的动手学习科学的地方之一是在实验室，但少量的实验经费和书本总不能达到这个目标。

"对很多学生来说，很多难以理解的概念都是在教材中介绍的，通过阅读去了解的，他们很难明白这些概念到底是什么意思，"（美国）谢南多厄河谷州立学校的校长琳达·考雷说。现在一个新颖的虚拟（现实）的网站帮助学生更好地掌握这些基本的科学概念。

"我是一个视觉学习者——对我来说，记住一张照片或一张图要比记住一段字母容易得多，"高中学生本·霍曼说。

"我们正通过教材学习这些概念，但接着我们用计算机、用另一种方法来了解这些概念，"另一个高中学生凯特·布莱克说。由维吉尼亚大学的工程师和学生开发的虚拟实验室，通过 50 个（虚拟）实验（如图 16.2 所示），结合教材和生动的动画解释了很多器件设备如半导体和发电机是如何工作的。

"虚拟现实是指我可以把不可见的东西显示出来，我们最常描述的是电场，你不可能看到它，但我可以用虚拟现实的手段使它们可以被看到，"维吉尼亚大学的工程师约翰·皮恩说。

网站定位在高中和大学一年级的学生——提供在线学习帮助，可以帮助学生完成回家作业和更好地直观地理解科学概念。

"在计算机屏幕上看，可以更好地记住那些画面，更好地理解这些概念，"凯特·布莱克说。（摘自一段网上的新闻，注意新闻的写作手法）

16.6.3 关于图像卡（或译做视频卡）

视频或图像电路，通常做成一个图像卡，有时也做在主板上，是为创建一个用显示器来显示的图像，对早期的只能处理文本的计算机来说显示文本是非常普通的任务。

随着多媒体和三维图像的应用越来越多时，图像卡的作用就变得很重要了，图像卡发展成一个高效率的处理电路，可以看成是 CPU 一个非常重要的合作处理芯片。到了（20 世纪）90 年代后期，图像芯片领域的开发速度已达到任何计算机其他技术领域不能超越的水平。发展的结果之一就是主芯片（这里应指 CPU）制造商和图像卡制造商的合并。

现在计算机图像卡由四个主要部分组成：
- 图像处理器
- 视频存储器
- 数字-模拟转换随机存储器（RAMDAC）

● 驱动软件

关于图像卡有两个主要参数。

1. 分辨率

分辨率常指可寻址能力，但更恰当的，它是指视频的锐度或细节描述能力，最初是显示器的一个参数，是由（扫描电子）束的尺寸和点数（有时用线数）决定的。一幅满屏图像由千万个像素组成，这时屏幕的分辨率用行数乘以列数确定的最大可显示像素来表示（如图 16.3 所示）。分辨率越高，可以显示越多的像素，所以屏幕上可同时显示更多的信息（图像越清晰）。

2. 色度

屏幕图像的每个像素是用三个不同颜色（红、绿、黄）的信号组合后显示的。每个像素的精细度由这三束光的强度以及另一个决定像素色度信息的存储量共同决定。用于存储每个像素信息的（二进制）位数越多，则图像的色彩越丰富。

要让眼睛感到看到全彩（图像），红、绿、黄各需有 256 个深浅不同的层次，即对三原色中的每一种都要有 8 位（一个字节）存储量，因此一个像素总的需要 24 位存储量。然而由于视频存储器的用法不同，实际上有些图像卡每个像素需要有 32 位存储量来显示真彩。

16.6.4 DirectX

随着新的芯片组、芯片组的新版本甚至是完全新的技术以惊人的速度不断出现，图像技术成为计算机工业中发展特别快的一个领域。这就给想要利用最新三维硬件的应用软件提出一个问题，因为任何应用软件开发商不可能为每个图像处理器编写源代码程序。

应用程序编程界面（API）就是用来解决这个问题的。API 的作用就是作为应用软件和它所用的硬件之间的中介。软件作者通过标准化的命令编程输出图形数据给 API，而不是直接给硬件。由硬件制造商编写的驱动程序则把这些标准代码编译成各硬件能运行的源代码格式。

DirectX 是 1995 年引入的一组编程工具集，可以帮助软件开发者创建在 Windows 平台的所有多媒体应用软件。它几乎覆盖了多媒体的所有方面，到了 DirectX 7.0 时代——1999 年引入的第六个主要版本，主要由以下部分组成：

Direct3D——用于实时三维图像。
DirectDraw——用于加速二维图像。
DirectSound——用于声音回放。
DirectPlay——用于网络连接（尤其针对互联网多人游戏）。
DirectInput——用于游戏操纵杆和其他相关部件（的输入）。
DirectMusic——用于基于信息的音乐数据（的输入）。

现在 DirectX 9.0 应用很广，并有适用于不同语言的多种版本。

Unit 17 User's Manual

Pre-reading

Read the following passage, paying attention to the question.
1) What can you do with MiraScan?
2) How can we scan reflective originals?

17.1 Text

Here is the Color Flatbed Scanner User's Manual; you may know how to write a product's manual after reading.

17.1.1 Introduction to MiraScan

MiraScan is the driver for your scanner. It is TWAIN compliant and designed to be user-friendly. With its iconlized user interface and fully logical taskflow design, you can complete a satisfactory scanning job with only a few mouse clicks. You can perform all of the following tasks using MiraScan:

① Preview, scan and import the reflective originals or transparencies into your image editing software.

② Adjust the quality of your scanned image before you actually start editing the image in your image editing software.

③ Apply batch scan functionality, allowing you to specify and switch among multiple scanning sessions in an image.

④ Create special effects to the scanned image by applying invert and mirror functions.

⑤ Use the Color Wizard to adjust the image easily and quickly.

17.1.2 Scanning Reflective Originals

Step 1 Place the original face down on the scanner glass plate. Note the direction of the original so that you will not scan the image in the wrong direction (Fig 17.1). Close the scanner lid.

Step 2 Open your application software.

Step 3 If this is your first time to scan, you may have to select the TWAIN source by choosing **Select source** in the **File** menu and then selecting **MiraScan** in your application software (You only need to do this once, unless you re-install your application software). Please note that the way to select the TWAIN

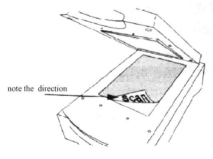

Fig 17.1   scanning reflective originals

source may differ according to the software you use. For details, please refer to the documents that come with your application software.

Step 4 Choose **Acquire** from your application software to bring up MiraScan (this may also differ according to the application you use).

Step 5 From the MiraScan main screens (Fig 17.2), select **Reflective** in the **Original** combo box, and then click **Preview**. A preview image will appear in the Preview Area.

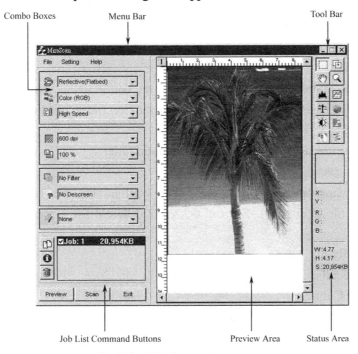

Fig 17.2　MiraScan main screens

Step 6 Adjust the scan area in the **Preview Area**.

Step 7 Use the options in the **Combo Boxes** to specify the Resolution, Scale…etc. That will apply to the scanned image.

Step 8 Use the options in the **Tool Bar** to adjust the image.

Step 9 If you need to add another scan area (scan job) to the original, push the **Duplicate** button in the **Job List** to add a new job (Fig 17.3). The repeat Step 6 and 7 to do the settings for that scan job.

Step 10 After you have finished with the settings for each scan job, press the **Scan** button to scan.

Step 11 A few seconds later, the scanned image(s) will be imported into your application software. You can start to edit the image(s).

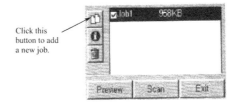

Fig 17.3　add a new job

17.1.3　Understanding MiraScan Functions

1. Menu Bar

There are three selectable menu items in the Menu Bar (Fig 17.2). Each menu item contains several options in its submenu.

2. File Submenu

MiraScan can record the settings you make for each scan session in a configuration file. With this feature, you can specify different settings for each scan job in each configuration file. Load the configuration file when you want to apply the settings. For example, you can save the settings you make for scanning magazine pictures in a configuration file and name it "Magazine". The next time you want to scan a magazine picture, you can load the "Magazine" configuration file and MiraScan will apply the settings recorded in it.

① Load Config.

Load the configuration file you created. When the configuration file is loaded successfully, MiraScan will apply the settings recorded in the configuration file automatically.

② Save Config.

Save current MiraScan settings into the configuration file.

③ Save Config As.

Save current MiraScan Settings as another new configuration file. Usually you can create a new configuration file in this way. You can designate different file names for each configuration file.

④ Reset Config.

When you select this item, the settings you make in MiraScan will be reset to its default value.

⑤ Auto Save Config.

When you select this item, the settings will be saved in a configuration file automatically as you exit MiraScan.

3. Settings Submenu

You can adjust several general scan settings in this submenu.

① Scan Size.

When you choose this item, the following dialog box (Fig 17.4) will appear.

In this dialog box, you can set the sizes of the areas to be scanned. First, choose the job number (the numbers of the scan areas listed in the Job List) from the job number list box and then set the size of the scan area, click in the **Scan Size** list box.

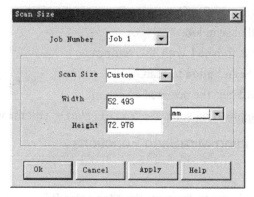

Fig 17.4 Scan Size dialog box

If you have set the original as Reflective (Flatbed) in the combo Boxes, you can choose the scan area size from the list. Four frequently used formats are listed: A4, A5, B5 and Letter. You can also customize the scan sizes by choosing Custom in the list box and then enter the desired width and height into the Width and Height text boxes. The default unit is inch. To change the unit, press in the list box with your mouse, and choose the desired unit from the list by clicking it.

If you have set the original as Transparent or Negative Film, the list box will contain only tow items: Max Area and Custom. Choosing Max Area, the full range of the image will be scanned. If you choose Custom, you can enter the sizes you want into the Width (max: 8 inches) and height (max: 10 inches) text boxes.

After you complete your settings, click on the OK button to confirm you settings and close the dialog box. You can then press the c button to close the dialog box without saving your settings or you can press the apply button to apply your settings to the image without closing dialog box.

② Monitor Gamma.

Generally, a monitor cannot display the real colors of an image in the real world. In order to have the monitor approximate the colors as possible, usually, you need to adjust the monitor gamma.

Adjustment of the monitor gamma can make the monitor approximate the colors of the original image more closely. Choose this item, and the following dialog box (Fig 17.5) will appear:

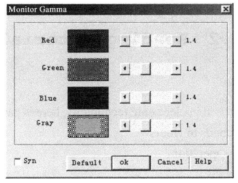

Fig 17.5　Monitor Gamma dialog box

From this dialog box, you can adjust the brightness in the midtone of display so that it can approximate the color details of the original images. Drag the scroll bars to adjust the gamma values of red, green, blue and gray until the monitor displays the original colors more accurately. If the Syn box is checked, you can adjust all the four gamma values simultaneously by dragging any scroll bar.

When the adjustment is done, press the ok button to confirm your settings and close the dialog box; otherwise, press the Cancel button to cancel your settings and close this dialog box. If you use the default value, click on the Default button and MiraScan will set the gamma value to 1.4.

（本文摘自某扫描仪的用户手册）

Technical Words and Phrases

approximate	[ə'prɔksɪmeɪt]	*adj*. 近似的，大约的　*v*. 近似，接近，约计
batch	[bætʃ]	*n*. 一批；一组；大量，批量
compliant	[kəm'plaɪənt]	*adj*. 顺从的，适应的

config	[kən'fɪg]	n. = configuration, [计算机专业]配置；组态，结构，布局；格局（软）
customize	[kʌstəmaɪz]	v. 定制，用户化
gamma	[gæmə]	n. 伽马（希腊语字母表的第3个字母Γ, γ），第三等
lid	[lɪd]	n. 盖子 vt. 给……盖盖子
original	[ə'rɪdʒɪn(ə)l]	adj. 最初的，原始的，独创的，新颖的 n. 原物，原作
reflective	[rɪ'flektɪv]	adj. 反射的，沉思的；熟虑的
settings	[setɪŋs]	n. 设置值，设定值
simultaneously	[sɪməl'teɪnɪəsly]	adv. 同时地
specify	['spesɪfaɪ]	vt. 指定，详细说明，列入清单
transparent	[træns'pærənt]	adj. 透明的，显然的，明晰的

for details	要得到详细资料，要详细了解
iconlized user interface	图标式的用户界面
mirror functions	镜像功能
negative film	底片的；负片的
scroll bar	滚动条

17.2 Reading Materials

17.2.1 NE555

NE555 is a timer usually used in electronic technology, here is part from its handbook.
- Timing From Microseconds to Hours
- Astable or Monostable Operation
- Adjustable Duty Cycle
- TTL-Compatible Output Can Sink or Source up to 200 mA
- Designed To Be Interchangeable With Signetics NE555, SA555, and SE555

Description

These devices are precision timing circuits capable of producing accurate time delays or oscillation. In the time-delay or monostable mode of operation, the timed interval is controlled by a single external resistor and capacitor network. In the astable mode of operation, the frequency and duty cycle can be controlled independently with two external resistors and a single external capacitor.

The threshold and trigger levels normally are two-thirds and one-third, respectively, of U_{CC}. These levels can be altered by use of the control-voltage terminal (CONT). When the trigger input (TRIG) falls below the trigger level, the flip-flop is set and the output goes high. If the trigger input is above the trigger level and the threshold input (THRES) is above the threshold level, the flip-flop is reset and the output is low. The reset (RESET) input can override all other inputs and can be used to initiate a new timing cycle. When RESET goes low, the flip-flop is reset and the output goes low.

When the output is low, a low-impedance path is provided between discharge (DISCH) and ground (GND).

The output circuit is capable of sinking or sourcing current up to 200 mA. Operation is specified for supplies of 5 V to 15 V. With a 5-V supply, output levels are compatible with TTL inputs.

The NE555 (Fig 17.6) is characterized for operation from 0 ℃ to 70 ℃. The SA555 is characterized for operation from –40 ℃ to 85 ℃. The SE555 is characterized for operation over the full military range of –55 ℃ to 125 ℃.

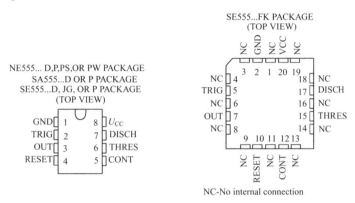

Fig 17.6　the package of NE555

The package and definition of pins of NE555 are shown in Fig 17.6.

（本文摘自某公司 NE555 芯片资料）

17.2.2　AD574A

AD574 is an A/D converter which converts analogy signal into digital signal, and usually used in electronic technology, here is part from its handbook.

1. Features (Fig 17.7)

- Complete 12-bit A/D Converter with Reference and Clock
- 8- and 16-Bit Microprocessor Bus Interface
- Guaranteed Linearity Over Temperature 0 ℃ to +70 ℃ – AD574AJ, K, L;
  　　　　　　　　　　　　　　　　　　–55 ℃ to +125 ℃ – AD574AS, T, U
- No Missing Codes Over Temperature
- 35 ms Maximum Conversion Time
- Ceramic DIP, Plastic DIP or PLCC Package
- Available in Higher Speed, Pinout-Compatible Versions (15 ms AD674B)

2. Product Description

The AD574A is a complete 12-bit successive-approximation analog-to-digital converter with 3-state output buffer circuitry for direct interface to an 8- or 16-bit microprocessor bus. A high precision voltage reference and clock are included on-chip, and the circuit guarantees full-rated performance without external circuitry or clock signals.

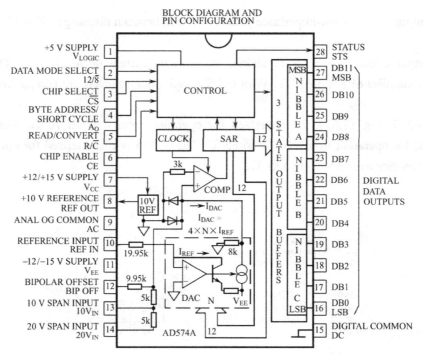

Fig 17.7   AD574A

The AD574A integrates all analog and digital functions on one chip. Offset, linearity and scaling errors are minimized by active laser-trimming of thin-film resistors at the wafer stage. The voltage reference uses an implanted buried Zener for low noise and low drift. On the digital side, 12L logic is used for the successive-approximation register, control circuitry and 3-state output buffers.

The AD574A is available in six different grades. The AD574AJ, K, and L grades are specified for operation over the 0 ℃ to +70 ℃ temperature range. The AD574AS, T, and U are specified for the –55 ℃ to +125 ℃ range. All grades are available in a 28-pin hermetically-sealed ceramic DIP. Also, the J, K, and L grades are available in a 28-pin plastic DIP and PLCC, and the J and K grades are available in ceramic PLCC.

（本文摘自某公司 AD574 芯片资料）

17.2.3   Cover letter

1. Purpose of a Cover Letter

A cover letter is a brief letter that introduces your resume to the company or organization to which you are applying. Your cover letter should emphasize why you want to work for that particular organization and why you would be a good fit. An effective cover letter engages the reader and encourages him or her to invite you for an interview.

A resume should always be accompanied by a cover letter. It serves as the first sample of your writing ability and attention to detail.

2. What to Include in your Cover Letter

① Formatting:

- Address your cover letter to a named individual, whenever possible.
- Make it brief - one page or less, with ample margins.
- Use an adequate font size – no smaller than font size 10.
- Include pertinent personal data: name, address, E-mail and phone number.
- Proofread for spelling or grammatical errors.

② Content:
- State why you are writing and for what position you are applying.
- Demonstrate energy and enthusiasm for the position.
- Highlight or expand on key information from your resume, but do not simply repeat what is listed.
- Actively sell your unique qualities and tell the reader why he or she should choose you.
- Target your skills, interests and experience to the needs of the organization.
- Show you have done your homework; emphasize why you want to work for that particular organization.
- Encourage the reader to take a closer look at your resume.

Here is an example of cover letter.

Date

Maxwell Mike
Ganco Engineering
4567 Main Street
Yokomo, IL 99999

Dear Mr. Mike:

Recently, I completed a cooperative experience with an engineering firm in Northern California where I was given the responsibility of managing a small project. I am looking for the opportunity to perform in this capacity for Ganco Engineering. I believe my background and experience will help me be an asset to your company in a very short time.

I have conducted survey and mapping assignments, participated in soil mechanics and foundation formation, and performed preliminary structural analysis and design for a grocery store. I was given the responsibility of producing a general lay-out for a new transportation system. I first surveyed the area, produced a traffic analysis and survey, researched city rules and regulations for compliance, and wrote a twenty-five page report depicting the feasibility of building a new freeway off-ramp. My mathematical skills are excellent and I am very conscientious about meeting deadlines and completing tasks unsupervised. If given the opportunity to accomplish an assignment, I can develop a plan that will meet the needs of the project.

I am very interested in becoming a part of this project after reading your company brochure and an article in the Civil Engineering Journal about your company's involvement in building a new mall in the area. I can be reached at the address and phone number below. I'll be calling your office within ten days to inquire on the status of my application. I look forward to hearing from you.

Sincerely,

Marilyn Appleton
12300 Hilltop Drive
Mantana, CA 99444
(919) 345-5566

17.2.4 Resume

A good resume is brief and concise, typed neatly in block form with no errors. It must be easy to read. When preparing your resume, cover the following points:

At the top of the page, put your name. On the next line place your mailing address followed by city, state, and zip. To the right of your name and address, place your home phone number. On the next line place your work phone number. Now you are ready to begin the body of your resume.

① In the first section of information, list your educational background. This section should be titled "EDUCATION". List the Degree and the year you received your degree on the first line. Second line should list the University your degree was received from and the city and state in which the University is located. If you received a Graduate Degree, you should list this prior to undergraduate work on one full line and list it in the same manner. If you have not fully completed your graduate work, list the amount you have completed.

② The second section on your resume should be your experience, and the grouping should be listed as "EXPERIENCE". Under experience should be military history, you should list all military positions in reverse chronological order.

③ Now, we want to explain the positions you have held. On the first line, list he dates when the positions were held and the title of the position. Then, in paragraph form, list the responsibilities of the position you held, the number of people you supervised and the type and amount of equipment for which you were responsible.

④ Listing your accomplishments and the positive impact these accomplishments made is important for an interviewer. An interviewer must know more about you than accomplishments you attain and the results and impact of those accomplishments. It is the results of you accomplishments, on which an interviewer will judge your job performance and effectiveness.

⑤ The third grouping on your resume should be special classes or training you received. This section should be titled "SKILL". To the right of the group title, list the name of the class/course and the date attended.

⑥ The fourth grouping on your resume should be involved honors/activities. This section should be titled " ACTIVITIES & HONORS ". You should list all honors, organizations, community service, offices held, etc.

Layout/format of resume is following:

Name (centered or left justified; bold and larger than other text)		
School address 1st line 2nd line Telephone E-mail address (E-mail address under your name looks fine also)	Set your table borders as "none" in Microsoft Word, so the borders will not appear on your resume document. The table format is solely to help you structure your resume and reformat easily.	Permanent Address 1st line 2nd line Telephone
Objective	Your concise objective here	

Education	• B.S. (Major), June 2006
	  Virginia Polytechnic Institute and State University (Virginia Tech) Blacksburg, Virginia
	  GPA: x.x/4.0	

Skills	First skill Another skill Etc.	
Experience	Employer Company/Organization, City & State location Job Title, Month/Year dates (DO use the bullet feature in MS Word to list each item) (DON'T try to manually insert bullets and manually add spacing to make your indentions line up)	
Activities & Honors	(Leadership position in organization) (Award) (Scholarship)	

17.3 Knowledge on Writing User's Manual（用户说明书写作知识）

使用说明书是随各种产品设备附带的书面材料，其形式有书本，小册子或散页之分，现在还有光盘，主要看它所说明的设备大小与复杂程度而定，其内容包括设备的安装、试验、维护等。

1. 特点

英语说明书在编写体裁等方面的特点是：

① 文句简短、扼要。

② 内容严格按照使用和维修时的先后顺序编排、划分类别章节，并给出醒目的标题及不同的序号。

③ 作参考用的附图和附表比较多，用以辅助文字的说明，直观性强。

④ 重要的零部件和操作程序均用大写字体标出。

⑤ 在操作与测试过程的说明中，祈使句的出现率大，被动语态也很常见。

⑥ 常用到很多组合词，翻译时注意与实际结合。

2. 说明书的编写

① 序——说明书的最前面往往有个序，用来简洁强调本产品的优点或特点，有时也说明本产品使用前应特别注意的事项，如关于安全使用，电视机、录像机等的电源要求，对所

接收信号的要求等，这一部分有时直接标明：本产品特点，注意事项等。

Warning: This VTR can be used with a power (mains) voltage of 100 to 110 V,115 to 127 V, 200 to 220 V or 230 to 250 V. 这句话说明了这个录像机的电源适用范围。

其主要用途是让用户在选购产品时可以一目了然地了解该产品的最主要特征及是否适用。还有常用的：Features（特点）、Precautions（保护措施）等。

② 目录（Contents）——列出说明书的说明内容的目录，便于查询。

③ 正文——对设备的各个部分进行描述和说明，如本单元课文及阅读材料所摘选的，一般是采用图文结合的方式，写作和翻译时都要注意对照。

如果有多项功能时，要注意逐项顺序介绍，重复步骤可写参见……文字要简洁。如课文中用：step1, step2…, 可以让使用者操作方便。

正文可以采用章节结构，也可以采用项目结构或列表结构，视具体需要说明的内容而定。在使用英语大写字母缩写时，第一次必须有全文表示，考虑到说明书（手册）的阅读对象并非一定是专业人员，使用专业术语最好也要用通俗的语言甚至图片加以说明。

17.4 Exercises

1. Put the Phrases into English

① 只要用鼠标点几下　　　　⑥ 重新安装应用软件
② 预览，扫描和输入　　　　⑦ 参考用户说明书
③ 批处理扫描功能　　　　　⑧ 扫描图像
④ 特殊效果　　　　　　　　⑨ 镜像功能
⑤ 盖上扫描仪的盖板　　　　⑩ 指定分辨率

2. Put the Phrases into Chinese

① driver program　　　　　⑥ list box
② combo boxes　　　　　　⑦ menu item
③ scroll bar　　　　　　　　⑧ submenu
④ tool bar　　　　　　　　　⑨ dialog box
⑤ preview area　　　　　　⑩ cancel your settings

3. Sentence Translation

① With its iconlized user interface and fully logical taskflow design, you can complete a satisfactory scanning job with only a few mouse clicks.

② Adjust the quality of your scanned image before you actually start editing the image in your image editing software.

③ Please note that the way to select the TWAIN source may differ according to the software you use.

④ MiraScan can record the settings you make for each scan session in a configuration file. With this feature, you can specify different settings for each scan job in each configuration file.

⑤ The default unit is inch. To change the unit, press in the list box with your mouse, and choose the desired unit from the list by clicking it.

⑥ These devices are precision timing circuits capable of producing accurate time delays or oscillation.

⑦ In the astable mode of operation, the frequency and duty cycle can be controlled independently with two external resistors and a single external capacitor.

⑧ The reset (RESET) input can override all other inputs and can be used to initiate a new timing cycle.

⑨ A high precision voltage reference and clock are included on-chip, and the circuit guarantees full-rated performance without external circuitry or clock signals.

⑩ The AD574A integrates all analog and digital functions on one chip.

4. Write a Document to Introduce the Menu Bar and How to Use the Notebook（Fig 17.8）

Fig 17.8　notebook main screen

（试写一个说明，介绍记事本的菜单及其用法）

17.5　课文参考译文

这里是彩色平板扫描仪用户手册，通过阅读你可以知道如何写一个产品的说明书。

17.5.1　MiraScan 介绍

MiraScan 是这台扫描仪的驱动程序，它是 TWAIN 专为用户设计的友好界面。通过它的按钮化用户界面和十分合理的任务流设计，你只要用鼠标点几下就可以完成全部的扫描工作。你可以用 MiraScan 完成下列任务：

① 预览、扫描和输入（反射式）原件（一般的文件，不透明的文件）或透明原件（照相底片等），提供给图像编辑软件。

② 开始用图像编辑软件编辑图像前，可以调整所扫描的图片的质量。

③ 用批扫描功能，可以让你在一幅图片上指定多个扫描区域，并开始多个扫描过程。

④ 通过反色和镜像功能可以在扫描的图片上创建特别的效果。

⑤ 用色彩向导可以很方便地调整图片的色彩。

17.5.2　扫描不透明的原件

第 1 步：把原件面朝下放在扫描仪的玻璃平板上，注意原件的方向（如图 17.1 所示），盖上扫描仪的盖板。

第 2 步：打开应用软件。

第 3 步：如果你第一次扫描，你可以在"文件"菜单中的选择扫描仪中选择 TWAIN 扫描仪（即本扫描仪），然后在应用软件中选择 MiraScan。（这一步你只需做一次，除非你重新安装你的应用软件）。注意根据你所使用的软件选择 TWAIN 扫描仪的方法可能有些不同。具体方法可以参考应用软件所附带的文件。

第 4 步：在你的应用软件中选择扫描新图可以打开 MiraScan（这一步也会因所用的应用软件的不同而不同。）

第 5 步：从 MiraScan 主界面（如图 17.2 所示）的源组合框中选择反射式（不透明），单击"预览"按钮，在预览区中会出现一幅预览的图片。

第 6 步：在预览区域内调整扫描面积。

第 7 步：用组合框中的选项指定分辨率、比例等，对扫描的图像设置扫描要求。

第 8 步：用工具条中的选项调整图像。

第 9 步：如果你需要再加上另一块扫描面积（扫描任务），按下"重复"按钮，在任务列表中加上一项新的任务（如图 17.3 所示）。重复步骤 6 和步骤 7 设置扫描任务。

第 10 步：在完成每个扫描任务的设置以后，单击"扫描"按钮进行扫描。

第 11 步：几秒后，扫描的图像就输入到你的应用软件中，你可以对图像进行编辑了。

17.5.3 MiraScan 功能

1．主菜单条

有三个可选的菜单项在主菜单条上（如图 17.2 所示），每个菜单项的子菜单含有一些选项。

2．文件子菜单

MiraScan 可以把你设置的每个扫描过程记录在一个设置文件中，利用这个特点，你可以为每个设置文件中的各项扫描任务指定不同的设置。当你想用这些设置时，调用相应的设置文件。例如你可以存储你扫描杂志图片的设置文件，并命名为"杂志"，下次你想要扫描杂志图片时，你可以调用"杂志"设置文件，MiraScan 会采用它所记录的设置方法。

① 载入设置。

载入你所创建的设置文件，当设置文件被成功载入后，自动采用设置文件中记录的设置方法。

② 保存设置。

把当前的 MiraScan 设置保存成一个设置文件。

③ 另存设置为。

把当前的 MiraScan 设置另存为一个新的设置文件。通常你可以用这种方法创建一个新的设置文件。你可以为每个设置文件指定不同的文件名。

④ 重新设置。

当你选择了这个选项，你在 MiraScan 中的设置将被重新设置成默认值。

⑤ 自动存储设置。

当你选了这个选项时，设置会在你退出时自动存储成一个设置文件。

3．设置子菜单

用这个子菜单可以调整一些扫描设置。

① 扫描尺寸。

当你选择这个项目时，出现下列对话框，如图 17.4 所示。

在这个对话框中，你可以设置要扫描的面积大小，首先，从任务列表框中选择任务号（扫描面积的编号列在任务列表中），然后选择扫描面积的尺寸，在扫描尺寸列表框中单击"选择"按钮。

如果你已经在（主界面的）**组合框**（原件选择框）中设置了原件为反射式（平板式），你可以从列表中选择扫描面积。那里列出四种常用的格式：A4, A5, B5 和信函。你也可以通过在列表框中选择自定义来自定义扫描尺寸，然后在宽度和高度的对话框中输入想要的宽度和高度，默认的单位是英寸。为了改变单位，用鼠标点击列表框的（箭头），在列表框中单击想要的单位。

如果你已设置了原件为透明的或照相底片，列表框中将只有两项：最大面积和自定义。选择最大面积，扫描图像的最大范围（到一张最大的图片）。如果你选择自定义，你可以在宽度（最大 8 英寸）和高度（最大 10 英寸）对话框中输入你想要的尺寸。

在完成设置后，单击"OK（确定）"按钮确定你的设置并关闭对话框。你也可以单击"取消"按钮，不保留你的设置并关闭对话框。或者你可以单击"应用"按钮对图像应用你的设置但并不关闭对话框。

② 显示器 γ 值（灰度系数）。

通常，一个显示器并不能显示在真实世界中一幅图片的真正色彩。为了使显示器显示的色彩尽可能与真实图片接近，需要调节显示器的 γ 值。

调节显示器的 γ 值可以使显示器显示的图像更接近原件的色彩，选择这一项，出现下列对话框，如图 17.5 所示。

从这个对话框中，你可以调节亮度到显示器的中等色调，从而使它更接近原件的色彩。拖滑动块可以调节红色、绿色、蓝色和灰度的 γ 值，使显示器更精确地显示出原件的色彩。如果同步框被点击（打勾），你可以通过拖动其中任何一个滑块来同时调节四个 γ 值。

当调节好以后，单击"确定"按钮确定你的设置并关闭对话框。否则单击"取消"按钮，不保留你的设置并关闭对话框。如果你单击"默认"按钮，扫描仪自动设置四个 γ 值均为 1.4。

17.6 阅读材料参考译文

17.6.1 NE555

NE555 是电子技术中常用的定时器集成芯片，这里从其使用手册中摘录了一部分：
- 定时器，定时范围从微秒到数小时
- 非稳态（中文中也称双稳态）或单稳态输出
- 占空比可调
- 与 TTL 兼容的输出，吸收（灌电流）和输出（拉电流）可达 200 mA
- 可与 Signetics（公司）的 NE555, SA555, SE555 互换（使用）

（性能）描述

NE555 是精密定时器电路，可产生精确的时间延时和振荡。在用于延时或称单稳态模式工作时，定时间隔是由一外接的电阻电容电路控制的，在非稳态（振荡）模式，其输出频率和占空系数可分别用两个外接电阻和一个外接电容来控制。

门槛电平和触发电平一般分别是 U_{CC} 的 2/3 和 1/3。利用控制电压端（CONT）也可以改变门槛电平和触发电平。当触发端（TRIG）输入电平低于触发电平，触发器被置 1，输出高电平。如果触发输入高于触发电平且门槛端（THRES）输入也高于门槛电平，则触发器清零，输出低电平。清零（RESET）输入端可以不管任何其他端的输入情况（将输出清零），常用于定时器一个新的工作循环的初始化。当清零端低电平时，触发器被清零，输出为低电平。当输出是低电平时，在放电端（DISCH）和地（GND）之间有一条低阻抗通路。

输出电路是可以接收灌电流或输出拉电流高达 200 mA。电源电压可以是 5~15 V，当用

5 V 电源时，输出与 TTL 芯片的输入兼容（即可以直接相接）。

NE555（如图 17.6 所示）的工作温度范围是 0 ℃～70 ℃，SA555 是–40 ℃～85 ℃，SE555 是军用芯片，其工作温度范围是–55 ℃～125 ℃。

图 17.6 给出 NE555 的封装及引脚定义。

17.6.2　AD574A

AD574 是一种模数转换芯片即把模拟信号转换成数字信号的芯片，是电子技术中常用的芯片，这里从它的说明书中摘录了一部分。

1. 特点（如图 17.7 所示）

- 自带参考电压和时钟的 12 位 A/D 转换器
- 8 位或 16 位微处理器总线接口
- 在下列温度范围确保线性度：0 ℃～+70 ℃——对 AD574AJ, K, L 类型
  　　　　　　　　　　　　　　–55 ℃～+125 ℃——对 AD574AS, T, U 类型
- 在工作温度范围内没有误码
- 最长转换时间 35 ms
- 陶瓷双列直插、塑料双列直插或 PLCC（塑料有引线芯片载体，一种封装形式）封装
- 另有与它的引脚输出兼容的高速类型（15 ms AD674B）

2. 产品性能描述

AD574 是 12 位连续模–数转换器，带有三态输出缓冲电路，可以直接与 8 位或 16 位微处理器总线相接。芯片中自带高精度的参考电压与时钟，电路不需要外接电路或时钟信号就能完成全部模–数转换功能。

AD574 在一个芯片上集成了全部模拟和数字功能，通过有源激光微调晶片级的薄膜电阻使芯片的输出偏移，线性度和换算误差达到最小。参考电压用一个内置稳压二极管来降低噪声和漂移。在数字方面，内置连续转换寄存器，控制电路和三态输出缓冲器，可以输出 12 位数字量。

AD574A 有 6 个不同的等级，AD574AJ, K 和 L 级的工作温度在 0 ℃～+70 ℃，AD574AS, T 和 U 级的工作温度在–55 ℃～+125 ℃。各种等级芯片都有陶瓷密封双列直插 28 脚的封装、J, K 和 L 级还有塑料双列直插 28 脚和陶瓷 PLCC 封装。

（附注：图 17.7 是 AD574 的内部框图和引脚图。）

17.6.3　求职信

1. 写求职信的目的

求职信是一封写给你所申请职位的公司或部门，介绍你个人基本情况的短信。在求职信中应强调你为什么想在这个公司以及为什么你将是一个很合适的人选（或为什么你很适合这个公司（的岗位））。一封成功的求职信能说服读信的招聘人员想给你作一个面试。

求职信通常还附有一个简历，其作用是展示你的写作能力，吸引（招聘人员）去读你的简历。

2. 求职信中应写些什么

① 格式：

- 尽可能写给某个人（例如招聘人员）。
- 简短，一页，留有适当的页边距。
- 用适当的字体，最小 10 号字体。
- 包含申请人的个人信息：姓名、称呼、电邮地址和电话。
- 不要有拼写或语法错误。

② 内容：
- 阐述你想申请的职位和理由。
- 说明你的工作能力和工作热情是适合这个职位的。
- 突出说明或详细说明你简历中的关键信息，但不要简单地重复简历。
- 强调你的独特之处，告诉招聘人员为什么他应该选择你。
- 针对公司的需要强调你的技能、兴趣和经历。
- 强调为什么你想要在这个公司工作，显示出你已经对这个公司有所了解。
- 尽量使招聘人员看了求职信后想（进一步了解你，）去读你的简历。

以下是一封求职信，供参考：

日期
麦克斯韦·麦克（公司某职员）姓名
（公司）地址

麦克先生：
　　最近，我在北加里福尼亚的一个工程公司中管理一个小项目，有了一次合作经验。我希望有机会在 Ganco 工程公司施展这方面的才能，我相信以我的背景和经历能够在很短时间内成为公司的有用人才。
　　我做过测量，制订计划，学过土力学和基础结构，为一个杂货店做过初步的结构分析和设计。我曾负责设计一个新的交通系统。我先测量面积，做交通分析和调查，研究城市法规，写了一份 25 页的报告描述了建设一个新的高速公路的进出口坡道的可能性。我的数学是很优秀的，我很有时间观念，能自觉按时完成任务。如果给我机会去完成一个项目，我会做出一个满足项目需要的计划。
　　在读过你们公司的宣传册和在民用工程杂志上的关于你们公司将在某地建设一个新的购物商厦的文章后，我很有兴趣参加这个项目，下面是我的地址和电话，十天内我会打电话到你们公司，询问我的申请情况，希望能尽快收到你们的回信。

真诚地
（申请人）姓名
（申请人）地址

17.6.4　简历

　　一份好的简历应该是简单、精练的，用表格的形式打印，不要有错且容易读。在准备简历时，要含有以下的要点：
　　在页面的顶部写你的名字。接着是你的邮件地址、城市、州、邮编。在你名字和地址的右边，放上你的家庭电话，下一行是你的工作电话（或手机）。现在你开始写你的简历的正文了。

① 第一部分，列出你的教育背景。这一节标为"教育"，第一行列出你接受教育的年限和学位。第二行列出大学及其所在城市和州。如果你是研究生，则以相同格式先列出研究生的情况，再列大学的情况。如果你没有完成学业，则列出你已完成的部分。

② 第二部分是你的经历，标题为"经历"。在经历下面先写你从军的经历，以反时间顺序（最新的最先）列出你在军队中的所有职位。（有些国家有要求男青年当一年兵的法规，没有可不写）

③ 再说明你所担任过的职位。在第一行列出任职时间和职位，接着以段落的形式，列出该职位的职责、管理的人数和类型、设备数。

④ 列出你的成就和这些成就的实际效果（影响）。这一点很重要。招聘人员必须多了解你这个人而不是你获得的成就和这些成就的结果和影响，但列出成就的结果可以让招聘人员判断你的工作表现和效率。

⑤ 简历中的第三部分是你所受过的训练或上过的专项课程，标题为"技能"，列出你参加过的训练/课程和日期（即证书）。

⑥ 第四段是所获得的荣誉/活动。标题是"活动和荣誉"，可列出所有获得的荣誉、参加过的组织、社会服务、当过办公室助手等。

简历的格式如下：

名字（聚中或放在左边；加粗且字号比其他文本大一些）		
学校地址 第一行 第二行 电话 电邮信箱 （E-mail 地址也可以放在你名字的下面）	把表格的边框设为"无"，这样简历就没有边框了，这种表格形式可以使你的简历结构清楚。	永久地址（家庭地址） 第一行 第二行 电话
目标	你的求职目标（职位）	
教育	B.S. (Major), June 2006（理学士（专业），2006.6） Virginia Polytechnic Institute and State University (Virginia Tech) Blacksburg, Virginia GPA: x.x/4.0	
经历	• Employer Company/Organization, City & State location 　Job Title, Month/Year dates 　（用项目标记列出各条）	
技能	First skill Another skill Etc.	
活动和荣誉	（在组织中的领导职位） （奖励） （奖学金）	

Appendix A Reference Answers

Unit 1

1. Put the Phrases into English

① direct current circuits

② amplifier

③ analog electronics

④ semiconductor diode

⑤ transistor effect

⑥ microprocessor 或 microcontroller

⑦ electrical engineering

⑧ power engineering

⑨ telecommunications engineering

⑩ internal devices

2. Put the Phrases into Chinese

① 汇编语言

② 交流电路

③ 无源电路

④ 三相电路

⑤ 数字电子技术

⑥ 逻辑门

⑦ 三维虚拟图像

⑧ 计算机编程

⑨ （在大学里）主修

⑩ 高级编程技术

3. Sentence Translation

① 你愿意去参加一个没有音响放大器、没有大屏幕或灯光效果的流行音乐会吗？

② 这（电视的发明）应归功于英国工程师贝尔德·约翰·洛吉（John Logic Baird），他追随马可尼（Marconi）的足迹，想用与传送声音相同的方式传送图像。

③ 今天所说的电子技术实际上是在发现晶体管效应以后开始（发展）的。

④ 令人不可思议的是他的想法奏效了，并从此诞生了集成电路工业。

⑤ 这个课程模块介绍了在线性应用范围中的半导体器件的特征。

⑥ 当前计算机及微处理器在电子工业的各个领域中应用十分广泛。

⑦ 这个模块中安排学生对一个简单的微处理器编程完成工业上典型的控制任务。

⑧ 然后通过编程利用这些器件完成控制（系统）等操作。

⑨ 重点是如何应用编程技术解决工程应用的实际问题。

⑩ 电子技术专业将为毕业生打下一个牢固的基础，学生毕业后可以从事电气工程师的职业。

4. Read and Translate It into Chinese

① 电路分析是电气工程学科中最基本的课程，在其他学科中电路分析也是十分有用的课程，通过学习这门课程所获得的技能不仅在诸如电子技术、通信技术、微波技术、控制和能源技术等电气领域中十分有用，而且在其他（看起来不同的）领域中也要用到。

② 数字集成电路对现代社会的影响是显而易见的，没有数字集成电路，将不存在当代的计算机和信息技术革命。数字集成电路是计算机与信息技术发展中最重要、最关键的技术，因为一片集成电路就可以存储大量信息和进行复杂的计算处理，使得计算机和信息技术的发展成为现实。

Unit 2

1. Put the Phrases into English

① electrical components

② Ohm law

③ limit current

④ voltage divider

⑤ transistor biasing circuits

⑥ block DC current

⑦ store electric energy 或 store electrical energy

⑧ inductive reactance

⑨ dielectric insulating material 或 insulating material

⑩ AC resistance

2. Put the Phrases into Chinese

① 称为容抗

② 单位是欧姆

③ 防止器件烧掉

④ 对交流电流有阻抗

⑤ 用一个螺丝调节

⑥ 呈圆柱形式

⑦ 阻直流，但通交流

⑧ 改变电感

⑨ 由公式给出

⑩ 音频放大器

3. Sentence Translation

① 电阻常用做限流器，限制流过器件的电流以防止器件烧坏，电阻也可用做分压器，以减小其他电路的电压，电阻还可用在晶体管偏置电路中和作为电路的负载。

② 电阻器一般是线性器件，它的（伏安）特性曲线形成一条直线。

③ 如果你对电子技术颇有兴趣，建议学会"彩色条形码"电阻的识别方法。

④ 电容可以隔直流，但能通过充电和放电的方式通交流。

⑤ 有各种各样形状和尺寸的电容。

⑥ （可调电感）有一个强磁的圆柱状铁芯，可以通过调节进入线圈（增加电感量）或从线圈中出来（减

少电感量）来改变电感。

⑦ 电容可用于滤波、旁路信号、定时电路和音频调谐电路。

⑧ 当有电流流过电感器时，电感器周围就有电磁场，电感器是能以电磁场的形式暂时存储电磁能量的电子器件。

⑨ 这个感抗与电感量和交流电的频率有关。

⑩ 输入电压加在原绕组（第一绕组）上，则在副绕组（第二绕组）两端输出感应电压。

4. Write Main Clause of the Following Sentences（写出下列句子的主句）

① Function is the sinusoid.

② The loudness is a result.

③ The sum equals the sum.

④ Finding can be accomplished.

⑤ Sinusoidal circuits can be analyzed.

Unit 3

1. Put the Phrases into English

① general-purpose meter

② analog meter

③ reverse the test leads

④ mechanical adjust

⑤ measure　resistance

⑥ positive voltage

⑦ measure current

⑧ voltage amplitude

⑨ dual-trace oscilloscope

⑩ signal generator

2. Put the Phrases into Chinese

① 模拟万用表

② 扩展范围

③ 特殊仪表

④ 具有功能及范围选择旋钮

⑤ 呈现一幅电子图像

⑥ 显示电压波形

⑦ 在屏幕上出现

⑧ 相位关系

⑨ 例如，作为一个例子

⑩ 串联接入电路

3. Sentence Translation

① 万用表是一种通用仪表，能用来测量直流和交流电路的电压、电流、电阻，有的还能测量分贝（放大倍数）。

② 一个模拟万用表可以测量正电压和负电压，只需简单地对调一下两个测试笔或拔一下极性开关。

③ 当测量电流时，电路必须断开，插入万用表表笔使之与被测电路或元件相串联。

④ 一个波形的频率可以通过在水平方向数出波形一个周期的厘米值来确定，将这厘米值乘上时间/厘米控制钮的设定值就得到它的一个周期所需的时间。

⑤ 一旦设定了垂直单位（电压/单位格），就确定了是用衰减器还是用放大器把输入信号转换成适当幅度的电压信号。

⑥ 利萨如图形可用来显示两个同频率信号之间的相位关系和根据一个已知频率的信号求出一个未知频率的信号。

⑦ 有两种万用表：用指针在标准刻度上的移动来指示测量值的模拟万用表和用电子数字显示器显示测量值的数字万用表。

⑧ 如果两个信号是同频率的，在示波器上就出现一个圆。

⑨ 通常，图像显示（电压）信号如何随时间变化：其垂直轴 Y 表示电压，水平轴 X 表示时间。

⑩ 所有的信号发生器都有一个频率范围开关，一个精调控制用来选择一个特定的频率，一个幅度控制用来改变输出电压的峰–峰值（幅值）和一些输出端口。

4. Translation

这本书应归类于电子技术类书籍，书中讲述了电子技术的基本理论、元件、器件、电路及电子系统，这本书不仅通俗易懂，而且还有许多电路基本测试和检修的切实可行的方法。

Unit 4

1. Put the Phrases into English

① PN junction

② anode

③ external resistance

④ bipolar transistor

⑤ a small current controls a large current

⑥ base current

⑦ enhancement MOS transistors

⑧ p-channel

⑨ integrated circuits

⑩ electron and hole

2. Put the Phrases into Chinese

① 半导体材料

② 正向偏置

③ 取决于外部电路的电阻

④ 过高的反偏电压

⑤ 是正比于基极电流

⑥ 几乎可看成是短路

⑦ 引起晶体管电路的稳定性（不好）问题

⑧ 数字技术

⑨ 最普遍的技术（最常用的技术）

⑩ 用两种互补型的晶体管——N 沟道（MOS 管）和 P 沟道（MOS 管）

3. Sentence Translation

① 这时二极管的内部电阻是很小的，将有一个较大的电流流过二极管，流过电流的大小取决于外部电路的电阻。

② 现在用的三极管大多数是 NPN 型，因为这种类型的硅管比较容易制作。

③ 缩写词 MOS 表示 M（金属）-O（氧化物） S（半导体），过去分别用金属做门极，氧化物做绝缘层，半导体做沟道、基底等的材料。

④ 电容值与你所选取的电路有关，并与你要放大的信号频率有关。

⑤ 如果你想要简单方便的放大电路，这款三极管电路虽然比较老却比较好。

⑥ 当然这样利用三极管去控制一个灯看起来毫无意义。

⑦ 这个特点也许很重要：如果开关只允许流过小电流，就可以用一个小（额定电流）开关去控制一个相当大电流的负载。

⑧ 当三极管中没有电流流过时，称三极管处于截止状态（完全不通）。

⑨ 请注意实际点亮灯的能量来自于示意图右面的电池。

⑩ 现在意义应该很清楚了，一个小小的直流电流源可以用来使三极管导通，直流电源的电流与灯泡中所流过电流相比是很小的。

4．Translation（注意粗体字的作用）

① 在这种情况下，内部电阻很大，导致只有一个极小的电流（取决于二极管的漏电流）流过二极管。

② 要使一个 NPN 型的双极型晶体管导通，基极必须加略高于发射极的正向电压（约为+0.6V）。

③ 目前，实现（设计）微电路中最普遍的技术是采用 MOS 管。

④ 两个电路的输入和输出端都需要电容来隔直流通交流。

⑤ 可以只用单一电压源供电，需要用两个等值电阻接在电压源和实际接地点中间，创建一个单独的直流接地点。

Unit 5

1．Put the Phrase into English

① electronic power supply 或 steady DC Voltage source

② bridge rectifier

③ pulsating DC

④ anode of diode

⑤ peak voltage

⑥ capacitor filter

⑦ charge and discharge

⑧ Zener diode

⑨ IEEE (Institute of Electrical and Electronics Engineers)

⑩ technical professional association

2．Put the Phrase into Chinese

① 设备的运行

② 把交流电转换成直流电的器件（设备）

③ 电源线

④ 根据（取决于）所需要的直流电压值

⑤ 半波整流器

⑥ 从而产生一个稳定的直流输出

⑦ 在电容的负极

⑧ 流过负载

⑨ 在加正向偏置（电压）的条件下

⑩ 一个串联（限制电流）电阻

3. Sentence Translation

① 然而在反向偏置时，除非外加电压达到稳压电压（U_z）（设计时确定的），否则它不导通。

② 从变压器副边流出的交流电流通过二极管构成的桥式整流器整流成为脉冲直流电。

③ 电容放电不久，另一个直流脉冲又到了，再次对电容充电使它达到峰值电压。

④ 组织的全称是电气电子工程师学会，虽然这个组织众所周知的名字为 IEEE。

⑤ IEEE 在电气工程、计算机和控制技术方面的出版物占世界出版文献的 30%。

⑥ 只有当停电时，当你在黑暗的房间里走动本能地想去开无用的开关时，你会发现电在你的日常生活中是多么重要。

⑦ 电从发电厂传到你家是通过一个很了不起的系统，这个系统称为电力（分布）网。

⑧ 通常提到的民用电就是单相、220V 交流电源。

⑨ 为了减少线路传输损耗，典型的长距离传输电压在 15.5 万伏到 75.5 万伏。

⑩ 它通常有断路器和开关，所以在需要时变配电站可以断开传输网的高压电力线或断开配电网的低压电力线。

4. Read Industrial Robots（5.2.2），Choose the Best Answer for Each of the Following

C A D B D

Unit 6

1. Put the Phrases into English

① Kirchhoff's Voltage Law

② voltage sources

③ the law of conservation of electric charge

④ at every instant of time

⑤ voltages across elements

⑥ radio transmission

⑦ frequency modulation

⑧ the frequency domain

⑨ linear resistor

⑩ amplitude modulation wave

2. Put the Phrases into Chinese

① 电流源

② 在这种情况下

③ 给出基尔霍夫第二定律

④ 引入"回路"的概念

⑤ KVL 的另一种表述法

⑥ 电压代数和

⑦ 正弦稳态响应

⑧ 日常用电（即民用电和一般工业用电）的电压

⑨ 时不变电路

⑩ 调制百分比

3. Sentence Translation

① 对电路的任何一个节点，在每一瞬时流入节点的电流和总是等于流出节点的电流和。

② 假设 $u(t)=1$ V，则材料顶部的电势（电位）比材料底部的电势高出 1 V。

③ 对任意给定的电压函数 $u(t)$，如果流过材料的电流 $i(t)$ 总是正比于给定电压函数 $u(t)$，则该材料被称为线性电阻。

④ 如果一个线性时不变电路的输入是正弦（信号），则其（输出）响应是同样频率的正弦波。

⑤ 我们可采用几种不同的分析方法，其中一种是：对某一瞬时，列出一组方程，方程组中的变量是电压，这种方法称为节点电压法。

⑥ 传送高频信号要比传送低频信号消耗的功率小，所以远距离传送包含信息的高频信号效率比较高。

⑦ 在示波器中可看到的一个调幅信号有调制峰值和调制谷底。

⑧ 在假设电路 A 部分中没有受电路 B 部分中变量控制的受控电源，反过来也一样（即假设电路 B 部分中没有受电路 A 部分中变量控制的受控电源。）

⑨ 求开路端电压 u_{OC} 的方法是从电路 A 旁边移走电路 B，求出节点 a 和 b 之间的电压就是 u_{OC}。

⑩ 假设给定的任意电路中含有若干个下列元件：电阻、电压源、电流源。

4. Translation（注意 it 指什么）

① 电阻用来限制流过器件的电流，以防止器件被烧掉（device）。

② 电感是一种电子器件，当有电流流过电感时，电感以电磁场形式暂时存储电磁场能量（inductor）。

③ 远距离传输含有信息的高频（信号）效率比较高（损耗小）（to...）。

④ 电感总要抵抗电流的变化，电感对直流电流没电阻，但对交流电流有阻抗（inductor）。

⑤ 让我们插入一个三极管来取代开关，显示三极管是如何控制流过灯泡的电子流的（transistor）。

Unit 7

1. Put the Phrases into English

① digital integrated circuit

② system design

③ application-specific integrated circuit (ASIC)

④ the productivity of chip designers

⑤ sophisticated design tool

⑥ fast response time

⑦ combinational logic block

⑧ valid signal

⑨ input-output delay

⑩ the highest frequency logic circuit

2. Put the Phrases into Chinese

① 超大规模集成电路

② 粗略地说

③ 在一个芯片上集成大量晶体管即集成电路

④ 做出全面的设计决定

⑤ 沟道宽度与沟道长度的比值

⑥ 组合块（电路）的输出锁存器

⑦ 用（计算机）高级语言定义（描述）

⑧ MOS 管的可靠性

⑨ 减小消耗功率

⑩ 实现电路的功能

3. Sentence Translation

① 摩尔定律指出：在一个芯片上器件集成的数量将每两年翻一番，这些年来（集成电路的发展）是符合摩尔定律的。

② 硅芯片的技术发展使得集成电路设计者可以在一个芯片上集成几百万个以上的晶体管，甚至现在可以在单一芯片上集成一个中等复杂的系统。

③ 在已知电路拓扑（结构）条件下，组合电路的延迟可以通过改变在电路中晶体管的规模来控制。

④ 但制作 PCB 板需要专用的设备，所以不推荐新手（初学者）用这个方法，除非已有现成的 PCB 板。

⑤ 通常（晶体管的尺寸）实际是指沟道宽度与沟道长度的比值。

⑥ 索尼公司和 DoCoMo 日本电话电报公司已达成协议建立一个合资公司开发一项基于装有索尼非接触 IC 卡的移动电话的新业务。

⑦ 通过这个平台，用户可以用他们的移动电话享受原先只能由 IC 卡提供的服务。

⑧ 这对我们来说是一件好事，因为伽马射线、X 射线和紫外线对生物有害。

⑨ 电子也可以在一个电视机的显像管中流动，在显像管中电子撞击显示屏，产生一个光（闪烁）点。

⑩ 这种技术的潜在应用可能包括电子票和网上信用卡业务，这些将给服务商和用户提供方便和增值。

4. Translation（注意 that 用法）

① 通常一个电子稳压电源就是把交流电转换成直流电的器件（设备）。

② 无论是何种方式拖动发电机，各种型号的（商用）电力发电机产生的都是三相交流电。

③（在未来，）你从竞争对手那里偷了一台笔记本电脑，却惊讶地发现在你的手里硬盘上的数据已经自我销毁了。

④ 变压器的任务是把 7 200 V 电压降为 240 V，成为家用标准电压。

⑤ 30 多年来，每隔 18～24 个月，微处理器的能力就翻一番。大多数观察家预计此趋势还将再延续 10 年左右。

Unit 8

1. Put the Phrases into English

① decimal system
② storage of information within the computer
③ printed or displayed
④ logic operation
⑤ combination of gates
⑥ provide an output high only when all inputs are high
⑦ above expressing
⑧ synchronous device
⑨ be combined with other gates
⑩ the versatility and flexibility of digital integrated circuits

2. Put the Phrases into Chinese

① 触发器
② 异步器件
③ 在任意时间
④ 当时钟脉冲信号来到时改变状态
⑤ 八进制和十六进制系统
⑥ 清除输入（信号）
⑦ 玩电视游戏（玩计算机游戏）
⑧ 上网（网上冲浪）
⑨ 用同一时钟脉冲（控制）转换
⑩ 用广义的定义

3. Sentence Translation

① 二进制系统是计算机的基本语言（系统）。
② 单个逻辑门在只需要单种逻辑功能时可以单独使用，它们也可以与其他门组合，以实现更复杂的逻辑运算。
③ 74HCT 系列是为使 74HC 系列与 74LS 系列（TTL）输入相匹配而设计的一种特殊系列，所以 74HCT 可以与 74LS 系列在同一电路中混合使用。
④ 一个寄存器是一组触发器，每个触发器能存储一个二进制位的信息。
⑤ 触发器保存二进制信息，门电路控制何时和如何把新的信息送入到寄存器中。
⑥ 一般来说，逻辑学是按推理的一般原则处理的科学。
⑦ 当计算机对所给的信息处理完毕后，输出的信息是以非十进制形式打印或显示出来的，因此这些输出信息也必须再转换，这次是转换回十进制系统。
⑧ 清除输入端在正常时钟控制时必须保持为逻辑 1。
⑨ 一个异步计数器中包含了若干个触发器，每个触发器的输出端与下一个触发器的输入端相连接。
⑩ 一个同步的计数器内部电路结构比较复杂，以保证所有的计数器输出都在同一个脉冲输入时改变。

4．Translations（注意 Which 的用法）

① 稳压管在反相偏置条件下，除非（外加电压）达到了所设计的稳压值，否则是不导通的。

② 一个现代的机器人，从效果上来看，是一个遥控操纵器，在机器人中用可编程的微处理器来取代人的大脑。

③ 摩尔定律，它预言在一个芯片上所集成的器件的数目将每两年翻一番，与这些年的（实际发展）很一致。

④ 通常这是由机器来做的，机器能做得非常快。

⑤ 大多数晶体三极管（BIJ）技术中是只用一种类型的晶体管（构成电路），MOS 管与它们不同，通常是用两种互补型的晶体 MOS 管——N 沟道和 P 沟道。

Unit 9

1．Put the Phrases into English

① make up a computer system

② high performance software

③ the input/output (or I/O) devices

④ CPU instructions reside in memory

⑤ store data to an output device

⑥ the control bus (或 control signal lines)

⑦ interrupt lines

⑧ read data from a memory

⑨ semiconductor integrated circuits

⑩ dynamic RAM memory

2．Put the Phrases into Chinese

① 影响系统的性能

② 通常与外部设备相连接

③ 从输入设备中读取数据

④ 与一台冯·诺依曼计算机的各种部件相连接

⑤ 数据总线

⑥ 80x86 系列微处理器（芯片）

⑦ 区分存储器的不同地址和不同的输入/输出器件

⑧ 这就提出了一个问题

⑨ 值得简要地提一下

⑩ 一个存储芯片的存储容量

3．Sentence Translation

① 对 CPU 而言，大量的输入/输出口就像存储器一样，因为 CPU 可以把数据存储在输出设备中，也可从输入设备中读入数据。

② 总线是指一组在系统各个部件之间传递各种电信号的导线。

③ 信号的传输速度非常快，以至于信号线即使只有几厘米的（传输）距离都有可能导致时间（不匹配）问题。

④ 当软件想要访问一些特定的存储位置或输入/输出设备时，它把相应的地址值放在地址总线上。

⑤ 主存是一个相当大且存取速度很快的存储器，用来存储 CPU 操作时的程序和数据。

⑥ 只要芯片与电源相连接，存储的信息始终有效。

⑦ 在冯·诺依曼结构中，如 80x86 系列计算机，所有的操作都在中央处理器中执行。

⑧ 静态 RAM 用起来方便且有较短的读和写周期（即读写速度快）。

⑨ 闪存最主要的用途之一是存储计算机的基本输入/输出系统（通常称做 BIOS）。

⑩ BIOS 软件有很多不同的作用，但它的最重要的作用是载入操作系统。

4．Translation

① 应当归功于英国的工程师约翰·罗杰克·贝尔德（John Logic Baird），是他追随马可尼的足迹，试着用与传输声音相同的方法来传送图像。

② 电感器是能以其周围电磁场的形式暂时存储电磁能量的电子器件。

③ 控制线是一组信号线，它控制处理器与系统其他部件的（通信）联系方法。

④ 如果想要实时显示快速变化的信号时，人们宁愿使用模拟示波器。

⑤ 由于机器人对其周围的环境反应不灵敏，所以需要这些昂贵的安装成本。

Unit 10

1．Put the Phrases into English

① with some similarities to…

② in an academic sense

③ structured language

④ the distinguishing feature

⑤ local variables

⑥ side effect

⑦ readable programs

⑧ high-level computer languages

⑨ assembly language instructions

⑩ comments describing the statements

2. Put the Phrases into Chinese

① 执行一个指定的任务

② 潜伏（隐藏）在一个程序中

③ 在各种情况下

④ 和程序的其他部分分开并隐藏起来

⑤ 给出汇编语言编程的替代方法

⑥ 与一条汇编语言对应

⑦ 用字母及数字符号

⑧ 使编程者容易书写（程序）

⑨ 花括号中的两条语句

⑩ 创建程序结构

3. Sentence Translation

① 模块-结构化语言的显著特征是代码与数据的分离。

② 函数是 C 语言的组成模块，在函数中可以进行各种编程。

③ 所有在这个函数中声明的变量只能在这个函数中应用。

④ 汇编语言指令用来描述由微处理器执行的每个基本操作。

⑤ 你甚至可以发现计算机正被许多旅馆用来处理订房，以前这一工作都是人工做的。

⑥ 微处理器是硬件部分，微处理器要通过执行一系列的称为软件的指令才能工作。

⑦ 微软为想要从事用微软产品进行开发的职业的人员提供了一些不同层次的认证项目（证书）。

⑧ 操作系统中最常用的是由微软开发的 Windows 系列操作系统。

⑨ 从计算机用于解决实际应用问题以来，软件的任务就是在应用的概念和计算机概念之间架起一座桥梁。

⑩ 目前首要的任务就是要对那些仍用昨天的软件技术来解决今天的软件问题的人进行培训。

4. Choices

A D B A C

Unit 11

1. Translate the Phrases into English

① still image

② relate to television

③ the most common way of displaying images

④ CRT or the cathode ray tube

⑤ pixel

⑥ create magnetic fields

⑦ move the beam horizontally

⑧ electron beam

⑨ emit red, green and blue light

⑩ electrical signals

2. Translate the Phrases into Chinese

① 把像点重新拼成一幅有意义的图像

② 第二个奇妙的特点

③ 撞击平坦的屏幕

④ 含有三个不同的部分

⑤ 三种颜色混合在一起

⑥ 卫星广播

⑦ 标准亮度的信号

⑧ 对……很熟悉

⑨ 有线电视节目

⑩ 选（电视节目）频道

3. Sentence Translation

① 要了解电视，先看一下人的大脑。

② 有时还可以看到液晶显示屏 LCD 和等离子显示屏，但与 CRT 相比它们还很少。

③ 有可以在真空管中产生电磁场的线圈。

④ 当电子撞击在磷上，会使屏幕发光。

⑤ 当电子束每次从左边扫到右边，为在屏幕上显示不同的黑、灰、白点，电子束的强度是变化的。

⑥ 一个彩色电视屏幕与黑白电视屏幕有三点不同。

⑦ 当彩色电视机要产生一个红点时，它射出红色电子束撞击红光磷。绿点和蓝点也同样。

⑧ 你可能对电视机可接收到的五种信号很熟悉。

⑨ 复合的电视信号是调幅的视频信号和独立的调频声音信号。

⑩ 但不幸的是这样很容易就可以偷取有线电视台的服务（即不付钱收看），所以电视信号以一种有趣的方式编码。

4. Translation

① 高清晰，菲利浦（为未来）提供最先进的产品

当朋友到我家时，他们将对我的新菲利浦数字电视机目瞪口呆，那真正的平面显像管和具有影院效果的清晰图像，真是无与伦比；菲利浦数字电视，世界第一。

② 是电视机，但不仅仅是电视机

看，我的菲利浦电视机，双重调谐的信号，能产生高品质图像的数字梳指滤波器，能得到最优化图像的视频器件，令人无法相信的（好听的）环绕立体声，所有的功能均可遥控，不，这不仅仅是一台电视机，这是我所想要的一切，并且更多。

Unit 12

1. Translate the Phrases into English

① digital camera

② light-sensitive device

③ advanced image sensor technology

④ compress the image

⑤ maximum image size

⑥ resolution of digital camera

⑦ another important factor

⑧ optical zoom

⑨ lossy compression

⑩ store images

2. Translate the Phrases into Chinese

① 传统的胶卷相机

② 不需要冲胶卷（过程）

③ 把光转换成一个电压

④ 调整对比度和清晰度（细节）

⑤ （在两个卷轴之间）卷胶片

⑥ 捕获图像的芯片
⑦ 图像质量有些损失
⑧ 观看实际图像的全景
⑨ 数字变焦
⑩ 存储卡

3. Sentence Translation

① 虽然原理可能与胶片照相机相似，但数码相机的内部工作情况完全不同。
② 数码相机的照片质量由这样一些因素决定：光学镜头的质量、捕获图像的芯片、压缩方法和其他部件。
③ 点数越多，分辨率越高，则更多的图像细节被捕获到（或译为照片更清晰）。
④ 还有一个限制的因素是图像的压缩方法，大部分数码相机采用压缩方法是为了能在给定的存储卡中存储更多的图像。
⑤ 例如用户可以选择黑白、底片效果、棕褐色（或称老照片）模式。
⑥ 很多第一代的数码相机中含有一或二兆的内存，可存30张标准质量的640×489像素的照片。
⑦ 虽然有用可移动存储器的趋势，数码相机仍可与计算机连接，下载照片。
⑧ 自拍也是数码相机都有的特点。一般是按下快门后延时10 s拍照。
⑨ 视频压缩是指尽可能多的舍去显示时无用的数据的技术。
⑩ 例如删去50%的输送数据可能只造成5%的视频信息损失。

4. Translation

近年来，专业集成电路的市场主要产品是逻辑门阵列和逻辑门阵列的改进产品——大规模逻辑门阵列（SOG）。SOG有很多优点，如快速转换时间、高集成度、高性能电路。随着更大规模的无沟道的SOG阵列的引入，传统的路由器可能不再能处理复杂度不断增加的VISI内部接口问题。

Unit 13

1. Translate the Phrases into English

① Internet-based communications
② local telephone services
③ communicate in real-time
④ create a kind of private chat room
⑤ contact family and friends
⑥ the transmission medium
⑦ be growing in popularity
⑧ be converted into data packets
⑨ electronic junk mail 或 spam
⑩ freeware and shareware

2. Translate the Phrases into Chinese

① 听起来很像普通的电话（服务）
② 在线服务
③ 用更新的杀毒软件（程序）
④ 扫描寄来及寄出的电子邮件（或译为扫描来往的电子邮件）

⑤ 在各地的参会者

⑥ 在视频摄像机前面

⑦ 提供远距离教学

⑧ 数字文字信息

⑨ 全球无线（通信）服务

⑩ 局域网和广域网

3. Sentence Translation

① 当然，没有一种技术是无缺点的，基于网络的通信也有很多缺点，如病毒、泄密和垃圾邮件。

② 互联网通信中发展最快的形式之一是即时消息（IM）（也可译做网上聊天）。

③ 对已经免费或固定计费上网的用户，网络电话软件提供了可打给世界各地的免费电话。

④ 网上通话由硬件和软件组成，使你可以用互联网作为电话的传输方式。

⑤ 因此要不断更新杀毒软件，它能扫描进出电子邮件中的病毒。

⑥ 开视频会议时，参会者必须用同样的客户软件或相匹配的软件。

⑦ 随着无线服务的发展，产生了多媒体信息服务（MMS），它可以传送包括文本、声音、图片和视频的信息给可接收 MMS 信息的手机。

⑧ 服务器有时被定义成网络上管理网络资源的一台计算机或设备。

⑨ 服务器这个词有双重含义，既可指硬件，也可指软件（这一点容易使人混淆）。

⑩ 每个小区域中由一个发射塔和一幢含有无线电设备的建筑物（后来的基站有多幢建筑物），组成一个基站。

4．Translation

MP3 或称为 MPEG 声音层 III，是一种压缩声音文件的方法，MPEG 是运动图象专家组缩写，这个专家组开发了包括 DVD 电影、高清晰度电视和数字卫星系统视频数据的压缩系统（标准）。（这是一种压缩比率较大的活动图像和声音的压缩标准。）

用 MP3 压缩系统可以减小一首歌的二进制数据量，但仍保持歌的声音接近 CD 质量。当你压缩一首歌时，歌的音质会有损失，这就是在一个较小的存储系统中携带较多的音乐文件所付出的代价。MP3 播放器可通过 USB 接口与计算机相连接，一个较小的（音乐）文件也可以使我们能较快地从网上下载一首歌（如图 13.6 所示）。

Unit 14

1. Put the Phrases into English

① absorb heat

② freezing temperatures

③ heat-exchanging coils

④ the expansion valve

⑤ low-pressure zone

⑥ air conditioner

⑦ compresses cool Freon gas

⑧ a magnetron control circuit

⑨ positive charge

⑩ sugar molecules

2. Put the Phrases into Chinese

① 冰箱的基本概念（原理）

② 产生冷的感觉

③ 在低温下蒸发

④ 在冰箱内部蒸发的液体

⑤ 流过扩张阀

⑥ 对人有毒性（或译为容量使人中毒）

⑦ 流过一组线圈（或译为流过一组散热管）

⑧ 从微波束中吸收能量

⑨ 扭来扭去（或来回扭动）

⑩ 水分子的共振

3. Sentence Translation

① 因为水蒸发时吸收热量，带来凉爽的感觉。

② 如果你把冰箱中的制冷剂放在皮肤上（绝对不是好主意，别这样做），当它蒸发时会使皮肤结冰。

③ 纯氨气是有毒的，如果制冷液泄漏对人是一种威胁（或译为是很危险的）。

④ 氟里昂液体流过扩散阀，在这过程中蒸发成冷的低压氟里昂气体。

⑤ 微波加热对液态的水最有效，对脂肪、糖、结冰的水加热效果就差一些。

⑥ 随着无线计算机网络日益普及，微波干扰已成为无线网络工作中值得关注的问题。

⑦ 第一种故意用微波去加热的食品是爆玉米，第二种食品是鸡蛋（鸡蛋在一个实验者面前炸开了）。

⑧ 到1970年后期，技术已得到很大的改进，价格迅速下降。

⑨ 真空吸尘器也许看起来像一个复杂的机器，但传统的真空吸尘器实际上仅由6个基本部件组成。

⑩ 可以把真空吸尘器的垃圾袋放在进风口和排气口之间的任何地方，只要空气流可以经过袋子。

4. Translation

防火墙本质上是你的计算机网络和互联网之间的一个保护系统。如果正确运用的话，防火墙可以阻止任何未授权的使用和访问你的计算机网络。防火墙的任务是根据你的设置仔细分析流入和流出网络的数据。它（忽视）不允许来自不可靠的、未知或值得怀疑的地址的信息进入（你的计算机）。防火墙在任何网络中都起到一个重要的作用，因为它对来自外部世界的各种侵袭起到保护和屏障的作用。

Unit 15

1. Put the Phrases into English

① computer keyboard

② be connected up as a matrix

③ cursor control keys

④ help information

⑤ letter keys

⑥ the natural evolution

⑦ opto-electronic mouse

⑧ firmware program

⑨ communications protocol
⑩ inkjet printer

2. Put the Phrases into Chinese

① 在商业环境中
② 加快数据的输入
③ 功能键和控制键
④ 基于 Windows 操作系统的图形界面
⑤ 来自传感器的信号
⑥ 一个装在（键盘）主板上的数字信号处理器
⑦ 具有照片质量（的图片）的输出
⑧ 激光打印机
⑨ 静电荷
⑩ 基本原理

3. Sentence Translation

① 计算机键盘是一组排列的按键（开关），按下每个按键时送给计算机一个相应的信号。
② 键的一个重要的参数是力-位移曲线，它显示了按下一个键需要多大的力和在键按下的过程中这个力是如何变化的。
③ 随着计算机在商业环境中应用的增加，需要提高数据输入的速度。
④ 在鼠标内部，相应每个按键都有一个开关和一个微处理器，微处理器分析来自传感器和开关之间的信号，用它的固化程序把信号转换成打包的数据送给计算机。
⑤ 喷墨打印机是在纸上喷出特别细小的墨点来产生一幅图像。
⑥ 激光打印机与复印机的工作方式类似，区别只是光源不同。
⑦ 粘好粉末图案后，鼓在下面的传送带送来的一张纸上滚一圈。
⑧ 因为触摸屏很容易使用，这些地方对新的雇员进行全面培训的时间就可以缩短。
⑨ 在繁忙的商店、旅馆的快速服务、交通运输中心等处的自服务系统触摸屏终端可用于改进对顾客的服务。
⑩ 最初，大部分激光打印机是单色的（在白纸上打黑字）。

4．Translation

所有的视频游戏通常都有的核心部件列表如下：
- 用户控制界面
- CPU
- 随机存储
- 软件内核
- 存储游戏的媒体
- 视频输出
- 音频输出
- 电源

玩家通过用户控制界面控制视频游戏，如果没有用户控制界面，视频游戏就成了被动的媒体，像有线电视。早期的游戏用操纵盘或操纵杆（控制），但现在大多数系统用带有各种按钮和特别功能的很复杂的控制器（即控制功能增加了）。今天视频游戏中所用的两种最通用的存储技术是 CD 和 ROM 存储条，现在的系统也提供一些固态的记忆卡，用来存储游戏和个人信息。较新的系统，如 PS2（一种游戏机型）有 DVD 驱动装置（即可用 DVD 存储）。

Unit 16

1. Put the Phrases into English

① slide
② business presentation
③ encyclopedia
④ find the desired information
⑤ provide feedback
⑥ driving simulator
⑦ education and entertainment
⑧ sales representative
⑨ vivid colours
⑩ 3D animations

2. Put the Phrases into Chinese

① 建立一系列的事件
② 多媒体资源
③ 最经济和最有效的方法
④ 技术和软件最适当的结合
⑤ 手提式（便携式）摄像机
⑥ 数字视频
⑦ 善于通过联想进行学习
⑧ 有纯娱乐的一面
⑨ 利用多媒体的先驱者（最先利用多媒体的）
⑩ 声音与视频的集成（结合）

3. Sentence Translation

① 多媒体是指结合两种或多种媒体（元素），通常有声音和视频支持，达到生动地表达一种思想（概念）的作品。
② 从此以后，软件和硬件开发商们开始争着把各种媒体形式集成到个人计算机中。
③ 如果能更快地实现声音的交互和音频（质量）的改进，则人和计算机之间的交流将变得更加生动。
④ 这种方法（称为数字视频）可在计算机上对所有的视频数据进行处理——包括编辑和产生特殊效果。
⑤ DVD 的直径、厚度与 CD 相同，它们用一些与制作 CD 相同的材料和制作方法进行制作。
⑥ 令人惊奇的是这些数据轨道细得无法使人相信，两个轨道之间只有 740 纳米（一纳米是十亿分之一米）。
⑦ 为了存储更多数据，DVD 已发展到有四个存储层，每面两层。
⑧ 视频或图像电路，通常做成一个图像卡，有时也做在主板上，是为创建一个用显示器来显示的图像。
⑨ 旋转 DVD 盘的驱动电动机：根据所读的轨道，驱动电动机被精确控制在每分钟 200 转～500 转。
⑩ 随着新的芯片组，芯片组的新版本甚至是完全新的技术的以惊人的速度出现，图像技术成为计算机

工业中发展特别快的一个领域。

4. Translation

二眼插头与三眼插头的区别是什么？

先看一下墙上的插座。你看到美国的 120 V 的标准插座（如图 16.4(a)所示）和 220 V 的中国插座（如图 16.4(b)所示），有两个垂直的槽，中间的下方有一个圆孔，左边的槽稍比右边的大一些，左边的槽为"零线（或中线），右边的槽为火线（相线），圆孔为地线。插头正好与这些插座匹配。

如果看一下你家里，你会发现几乎所有带金属外壳的家电都用三孔插座。还有些外壳是塑料的也用三眼插座，例如你的计算机，其机箱内部有一个金属外壳的电源。接地是指外壳直接接在"地"线上，是为了防止人们在用金属外壳的家电时触电。

假设在不接地的带金属外壳的设备内部有一根导线松动了，碰到了金属外壳，如果正好是相线，则金属外壳现在带电，当人碰到外壳就触电了。外壳接地后，从松动的相线上流出的电流直接流入地，这时保险丝盒里的保险丝烧掉了，家电不工作了，但不会伤到人。

如果你断开接地插头或用一个便宜的二眼插座（接线板），把一个三眼插头的家电插到二眼的插座中，会怎么样呢？可能家电仍可以用，但当你这样做时，你把如果电线松动造成外壳带电后（三眼插座）对你的保护措施去掉了。

Unit 17

1. Put the Phrases into English

① with only a few mouse clicks

② preview, scan and import

③ batch scan functionality

④ special effects

⑤ close the scanner lid

⑥ re-install application software

⑦ refer to the User's Manual

⑧ scan the image

⑨ mirror functions

⑩ specify the resolution

2. Put the Phrases into Chinese

① 驱动程序

② 组合框

③ 滚动条

④ 工具条

⑤ 预览区域

⑥ 列表框

⑦ 菜单条

⑧ 子菜单

⑨ 对话框

⑩ 取消设置

3. Sentence Translation

① 通过它的按钮化用户接口界面和完全附合逻辑的任务流设计，你只要用鼠标点几下就可以完成全部的扫描工作。

② 在实际开始用图像编辑软件编辑图像前调整所扫描的图片的质量。

③ 注意根据你所使用的软件选择 TWAIN 扫描仪的方法可能有些不同。

④ MiraScan 可以把你设置的每个扫描过程记录在一个设置文件中，利用这个特点，你可以为每个设置文件中的各项扫描任务指定不同的设置。

⑤ 默认的单位是英寸。若要改变单位，可以用鼠标点击列表框的（箭头），在列表框中单击想要的单位。

⑥ 这些器件是精密定时器电路，可产生精确的时间延时和振荡。

⑦ 在非稳态（振荡）模式，其输出频率和占空系数可用两个外接电阻和一个外接电容来独立控制。

⑧ RESET 输入端可以不管任何其他端的输入情况（将输出清零），常用于定时器一个新的工作循环的初始化。

⑨ 芯片中自带高精度的参考电压与时钟，电路不需要外接电路或时钟信号就能完成全部模–数转换功能。

⑩ AD574 在一个芯片上集成了全部模拟和数字功能。

4．Write a Document to Introduce the Menu Bar and How to Use the Notebook（Fig 17.9）

（略）

Appendix B Technical Vocabulary Index

A

		所在单元
aborigine	土著居民	16
absorb	吸收，吸引	14
AC	交流电（流）	1
accelerate	加速，促进	3
account	会计；账户	13
acronym	首字母简略词	4
adapter	多头电源插座	13
address	访问，地址	9
admittance	导纳（电阻的倒数）	6
affix	贴上；粘上	15
alcohol	酒精	14
algebraically	代数	6
alphanumeric	字母与数字混排的	10
alternate	交替的	5
ammonia	氨，氨水	14
amplifier	放大器	1
amplitude	振幅	3
amplitude modulation	（AM）调幅	6
analog	模拟	1
analogous to	类似于……	6
ADC	模-数转换器	12
AND	逻辑"与"	8
animation	动画	16
anode	阳极，正极	4
antenna	天线	11
applications	应用软件	1
application-specific	专用的	7
approximate	近似的	17
architecture	计算机的结构	9
array	排列，阵列	15
assembly	汇编	10
assembly mnemonic	汇编助记忆码	10
assets	资产，资源	16
assign	指定	9
asynchronous	异步的	8

· 259 ·

B

batch	一批	17
be affected by …	受……影响	2
be assumed to	被假设成……	4
be available	是可利用的	2
be connected to	连接到……	3
be essential for/to …	对……很必要的	7
be labeled	被标志为……	4
be proportional to	与……成比例	2
be used to …	被用于……	2
beam	光束,电波	3
biased	偏压,偏置	4
binary	二进制位的	8
bipolar	双极性的	4
break down	(二极管)烧坏	4
bridge(full-wave)rectifier	桥式全波整流器	5
buddy	搭挡	13
bugs	原因不明的故障	10
bundle	捆扎	13
bus	(计算机)总线	9

C

cable	电缆,有线电视	11
capable of V-ing	可以做……的	3
capacitance	电容量	2
capacitor	电容器	2
career	事业,速度	1
cartridge	盒式磁盘[带](机)	15
cathode	负极,阴极	4
CRT	阴极射线显像管	3
chamber	室,(枪)膛	14
channel	信道,频道	11
characteristic	特性,特征值	1
charge	电荷,充电	2
CCD	电荷耦合器件	12
chat	聊天	13
chip	电路芯片	7
chrominance	色度	11
circuit	电路	1

cityscape	都市风景	16
clerk	售货员；店员	15
client	委托人	13
code	代码，编码	8
collector	集电极	4
column	列	15
combinational logic	组合逻辑	8
commercially available	有成品供应的	5
communicates with …	与……通信	9
compartmentalization	把……分门别类，把……分成区	10
compensate	补偿	3
complementary	补充的	4
compliant	适应的	17
compress	压缩	12
compression algorithms	压缩格式	12
condense	（使）浓缩，精简	14
config	配置，结构	17
configure	设置，设定	8
consequence	结果，推论	6
consumer	消费者；用户	12
contrast	与……对比，对比度	12
coordinate	坐标（用复数）	16
corona	光环	15
covert…into	把……转换	11
current operated device	电流控制器件	4
customize	定制，用户化	17
cycle	循环；周期	9

D

DC	直流电（流）	1
debug	排除计算机程序中的错误	10
decibel	分贝	3
decimal	十进的，小数的	8
defect	缺点（故障，缺乏）	3
deflect	（使）偏斜，（使）偏转	3
delay	推迟；延缓，阻塞	7
depict	描述	6
dielectric	电介质	2
differentiate	区别，辨别，求……的微分	9
DSP	数字信号处理器	12

diode	二极管	4
discharge	放电器，放（电）	5
discrete	离散的；不连续的	10
displacement	取代，置换	15
dissipate	驱散	14
dissolve	渐渐消隐，溶化	16
distinct	独特的	12
document	公文，文档	16
documentation	文件管理，文件编制	10
dominant	占优势的，主要的	4
dot	点，在……上打点	11
drum	鼓	15
dual-trace oscilloscope	双踪示波器	3
dynamic	动态的	9

E

edge-triggered flip-flop	边缘触发式触发器	8
effect	效果，招致	16
electrical	电的，有关电的	2
electronic power supply	电子稳压源	5
electronic technology	电子技术	1
emitter	发射体，发射极	4
employer	雇主，老板	13
encyclopedias	百科全书式的	16
enhancement	增强（提高，放大）	4
evaporation	蒸发	14
excessive	过度的	10
exploit	开发	12
exposures	暴露	12
extend	扩大	8

F

facilitate	使便利	15
family	系列	9
filament	灯丝；细丝	11
film	胶卷，影片	12
filter	滤波器	5
firmware	固件	15
flexibility	灵活性，弹性	8
flip-flop	触发器	8

flowchart	流程图	7
focus	焦点，焦距	3
for details	要详细了解	17
formula	公式	2
reverse biased	反偏置	4
forward biased	正偏置	4
frame	框架	7
freon	氟利昂	14
FM	调频	6
fuser	熔辊	15

G

gain	增益，放大倍数	4
gamma	伽马（γ）	17
gate	门（电路），栅极	8
germanium	锗	4
give sb. credit for sth	为……赞扬[肯定]某人	9

H

hardware	（计算机）硬件	10
hexadecimal	十六进制的	8
horizontally	水平地	3

I

I/O port	输入/输出端口	1
iconlized user interface	图标式的用户界面	17
ignorant	无知的	9
image	图像，映像	1
impedance	阻抗	6
implement	工具，实现	7
implementation	系统开发的具体实现	7
impose	施加影响	10
in parallel with	与……并联	3
in series with	与……串联	3
in terms of	以……的观点	7
in various shapes and sizes	各种形状和尺寸	2
incorporate	合并，使组成公司	16
individual	单独的；特殊的	8
inductance	电感量	2
inductor	电感器	2

industry-standard JPEG format	工业标准格式	12
inkjet printers	喷墨打印机	15
instant	立即的	13
instruction	指令	10
instrument	工具,手段	1
insulating	绝缘的	2
integrate	集成	7
Integrate …into …	把……集成……	11
integrated circuit (IC)	集成电路	7
intensity	强度,亮度	11
internal resistance	内阻	4
interrelate	(使)互相联系	7
interrupt	中断,中断信号	1
inverter	反相器	7
isolation	隔离,绝缘	5

J

jack	插孔,插座	3
joule	焦耳	14

K

keep pace with …	与……保持一致	7
keyboard	键盘	15

L

lag network	滞后网络	6
Laser Printer	激光打印机	15
latch	寄存器	7
lead network	超前网络	6
leakage	漏,泄漏	4
lens	透镜;镜头	12
lid	盖子	17
liquid	液体	14
location	地点,位置	9
loop	回路	6
lossy	有损耗的,致损耗的	12
lubricate	润滑,加润滑油	14

M

magnetron	磁电管，磁控管	14
magnification	扩大，放大倍数	12
maintenance	维护	16
major in	（在大学里）主修	1
make sense	有意义	10
manipulate	操作，使用	16
manufacturer	制造商	12
matrix	矩阵，真值表	15
measure	测量	3
medicine	药	1
medium	媒体	12
membrane	薄膜，细胞膜	15
mesh	网络	6
meter	米，仪表	3
microcircuit	微电路	4
microprocessor	微处理器，单片机	10
microwave oven	微波炉	10
minimize	最小化	12
miraculous	不可思议的	1
mirror functions	镜像功能	17
mnemonic	记忆的，记忆码	10
moderate complexity	中等规模	7
module	模块	1
molecule	分子	14
monitor	监视器，监控	12
monochrome	单色画；单色照片	12
motherboard	（计算机）主板	9
mouse	鼠,鼠标（（复）mice）	15
multimedia	多媒体	16
MMS	多媒体信息服务	13
multimeter	万用表	2

N

N gate	"非"门	8
NAND	"与非"	8
narration	讲述	16
needle	针	3
negative	底片，负的	12

negative film	底片的；负片的	17
node	节点	6

O

octal	八进制的	8
opcode	操作代码	10
operand	操作数	10
operating system	操作系统	10
opto-electronic	光电的	15
OR	"或"	8
original	最初的，原始的	17
oscilloscope	示波器	3
outwit	瞒骗，以智取胜	16
overscore	字等上面或中间的线	8

P

package	打包，封装	7
panel	面板，仪表板	12
passive	无源的	1
peak-to-peak voltage	电压峰–峰值	3
period	周期	3
periodically	周期的	9
phase shift	相位漂移，移相	3
phasor	相量，相量图	6
phenomenon	现象	15
phosphor	磷	11
photography	摄影	12
physical limit	硬件限制	7
pin	引脚	7
pixel	像素	11
plasma	等离子显示器	11
polarity switch	极性开关	3
polysilicon	多晶硅	4
pop concert	流行音乐会	1
powder	粉，粉末	15
power supplies	电源	1
presentation	陈述，赠送	16
prime	最初，第一流的，根本的	12
print head stepper motor	喷头步进电机	15
privacy issue	隐私泄漏	13

probe	探针	2
program	程序	1
projector	放映机	16
prompt	提出，〈计〉提示符	9
protocol	草案，协议	15
pulsate	（脉等）搏动	5

R

raw	未加工的	12
reactance	电抗	2
rectify	整流，检波	5
refer to…as	把……称为	6
reference	参考书目，介绍信（人）	16
reflective	熟虑的	17
refresh	刷新	9
refrigerant	致冷剂	14
refrigerator	冰箱，制冷机	14
repetitive	重复，迭代	6
resistance	电阻值	2
resistor	电阻器	2
resolution	分辨率	12
respond to	对……响应	6
ripple	脉动，波动	5
Rms voltage	电压有效值	3
roller	辊，卷	15
row	排，行	15
RS232 serial port	RS232（协议）串行口	12

S

satellite	人造卫星	11
saturation	饱和（状态）	11
screwdriver	螺丝刀，改锥	7
scroll bar	滚动条	17
semiconductor	半导体	1
sensitive	敏感的，感光的	12
sepia	乌贼的墨，棕褐色	12
sequence	次序，顺序，序列	11
settings	设置	17
shield	防护物，护罩	14
shine	照耀，发光	15

short message service (SMS)	短消息服务	13
silicon	硅	4
simultaneously	同时地	17
sinusoid	正弦曲线，正弦	6
size	计算机可同时处理数据的二进制位数	9
slide	滑动，幻灯片	16
smooth	平滑，变平静	5
software	（计算机的）软件	10
solder	焊料，焊接，焊补	7
sophisticated	精密复杂的	7
soundtrack	音轨	16
source code	源代码	10
spam	垃圾邮件	13
specify	指定，详细说明	17
spontaneously	自发地，自生地	16
stability	稳定性	4
static	静态的，静力的	1
storage	存储，储藏（量），储藏库	12
style	风格	16
subcomponents	子部件	7
subroutine	子程序	10
substitute	替代物；代理人	8
substrate	基底，底层，下层	4
succession	连续，继承	11
superimposed	有层理的，重叠，双重	3
switch	开关，电闸，转换	3
symbol	符号，记号，象征	4
syn	同步，共同	17
synchronous	同时发生的，同步的	8

T

technique	技术，技巧，方法	1
temporary	暂时的；临时的	10
terminal	终端，接线端	4
the credit goes to…	归功于……	1
thrive	兴旺，繁荣	1
thumbnail	极小的东西	12
to serve as…	用做……	2
toggle	双态元件，触发器	8
toxic	中毒的，有毒	14

trade-off analyses	平衡分析	7
transformer	变压器	5
transient analysis	暂态分析	6
transistor	晶体管	1
transmission	传送	13
transparent	透明的	17
trigger	触发,扳机	16
twist	扭曲	14

V

vacuum	真空,真空吸尘器	11
valve	阀;活门	14
variable	变量,易变的	10
VCR	盒式录像机,磁带式录像机	11
vertically	垂直地	3
viable	可行的;可实施的	15
videoconference	视频会议	13
viewfinder	取景器	12
virtual	虚的,虚拟的	1
virus	病毒	13
VoIP	IP 电话	13
voltage	电压	2

W

well-informed	见闻广博的	7

Z

Zener diode	齐纳二极管,稳压二极管	5
zone	存储区,区域	14
zoom	图像电子放大	12

Reference

[1] Allan R.Hambley 著. 李春茂改编. 电子技术基础（英文改编版）[M]. 北京：电子工业出版社，2005

[2] Ken Martin. 数字集成电路设计（英文版）[M]. 北京：电子工业出版社，2002

[3] Leonard S.Bobrow. 线性电路分析基础（英文版）[M]. 北京：电子工业出版社，2002

[4] 俞光昀，王炜. 计算机专业英语[M]. 北京：电子工业出版社，2002

[5] 尤毓国. 英语科技情报文献阅读[M]. 北京：新时代出版社，1991

[6] 季键. 科技英语结构分析与翻译[M]. 北京：新时代出版社，1983

[7] 张叔方. 科技英语捷径[M]. 北京：北京语言出版社，1990

[8] Harry M. Hawkins. *Concepts of digital electronics*. Blue Ridge Summit, Pa.: Tab Books, 1983

[9] Fredrick W. Hughes. *Illustrated guidebook to electronic devices and circuits*. Englewood Cliffs, N.J.: Prentice-Hall, 1983